WIRELESS PERSONAL COMMUNICATIONS

COMMUNICATIONS

Channel Modeling and
Systems Engineering

THE KLUWER INTERNATIONAL SERIES
IN ENGINEERING AND COMPUTER SCIENCE

WIRELESS PERSONAL COMMUNICATIONS

Channel Modeling and Systems Engineering

edited by

William H. Tranter
Brian D. Woerner
Theodore S. Rappaport
Jeffrey H. Reed
Virginia Polytechnic Institute & State University

KLUWER ACADEMIC PUBLISHERS
Boston / Dordrecht / London

Distributors for North, Central and South America:
Kluwer Academic Publishers
101 Philip Drive
Assinippi Park
Norwell, Massachusetts 02061 USA
Telephone (781) 871-6600
Fax (781) 871-6528
E-Mail <kluwer@wkap.com>

Distributors for all other countries:
Kluwer Academic Publishers Group
Distribution Centre
Post Office Box 322
3300 AH Dordrecht, THE NETHERLANDS
Telephone 31 78 6392 392
Fax 31 78 6546 474
E-Mail <orderdept@wkap.nl>

 Electronic Services <http://www.wkap.nl>

Library of Congress Cataloging-in-Publication Data

Wireless personal communications : channel modeling and systems engineering / edited
by William H. Tranter ... [et al].
    p.cm.
  Includes bibliographical references and index.
  ISBN 0-7923-7705-2 (alk. paper)
  1. Personal communication service systems--Congresses. I. Tranter, William H.

TK5103.485 . W5623 1999
621.3845--dc21

                                     99-047414

Printed on acid-free paper.

Printed in the United States of America

This printing is a digital duplication of the original edition.

TABLE OF CONTENTS

PREFACE

The papers appearing in this book were originally presented at the 9th *Virginia Tech/MPRG Symposium on Wireless Personal Communications*. The *Symposium on Wireless Communications*, which is an annual event for Virginia Tech, was held on June 2-4, 1999. The 1999 symposium was co-sponsored by MPRG, the Division of Continuing Education, University International Programs, and the MPRG Industrial Affiliate Sponsors.

Much of the success of our annual symposium, as well as the success of MPRG's research program, is due to the support of our industrial affiliates. Their support allows us to serve the wireless community through research, education and outreach programs. At the time of the 1999 symposium, the MPRG affiliates program included the following organizations: Army Research Office, AT&T Corporation, Bellsouth Cellular Corporation, Comcast Cellular Communications, Inc., Datum, Inc., Ericsson, Inc., Grayson Wireless, Hewlett-Packard Company, Honeywell, Inc., Hughes Electronics Corporation, ITT Industries, Lucent Technologies, Motorola, National Semiconductor, Nokia, Nortel Networks, Qualcomm, Inc., Radix Technologies, Inc., Salient 3 Communications, Samsung Advanced Institute of Technology, Southwestern Bell, Tantivy Communications, Tektronix, Inc., Telcordia Technologies, Texas Instruments, TRW, Inc., and the Watkins-Johnson Company

As can be seen from the Table of Contents, the papers included in this book are divided into six sections. The first five of these correspond to symposium sessions, and cover the following topics: Propagation and Channel Modeling (4 papers), Antennas (6 papers), Multiuser Detection (3 papers), Radio Systems and Technology (4 papers), and Wireless Data (3 papers). The last section contains invited poster papers (2 papers).

The first group of papers deals with Propagation and Channel Modeling. The first paper, *Very Near Ground RF Propagation Measurements and Analysis*, by Thad Welch, Michael Walker, and Ray Foran, treats the propagation characteristics of a cordless phone antenna when the antenna is placed near the ground. A situation like this might exist if an incapacitated person, lying on the ground, has access to a cellular or cordless phone. Results of their study show that a significant decrease in signal strength (as much as 12 dB) can occur if a person using the phone falls from a sitting to a prone position. The second paper in this section, *Identification of Time-Variant Directional Mobile Radio Channels*, is co-authored by R. S. Thoma, D. Hampicke, A. Richter, G. Sommerkorn, A. Schneider, and U. Trautwein. Their paper describes a broadband channel sounder which allows a full statistical analysis of the Doppler-delay-azimuth statistic of mobile radio channels. The measurement procedure uses processing based on the ESPRIT algorithm. The third paper in this section is co-authored by B. L. Johnson, Jr., P. A. Thomas, D. Leskaroski, and M. A. Belkerdid, and is entitled *Propagation Measurements and Simulation for Wireless Communication Systems in the ISM Band*. They use both deterministic and stochastic models to study propagation coverage in the 2.4 GHz ISM band for an area in South Florida. The result of their study is a Hata-Okumura model implemented in Mathcad™. The results show that Mathcad™ is a practical tool for simulating propagation coverage. The next paper in this group was contributed by Dave Crosby, Steve Greaves and Andy Hopper. This contribution, entitled *A Theoretical Analysis of Multiple Diffraction in Urban Environments for Wireless Local Loop Systems*, studies the use of simulation to study multiple diffraction effects in wireless local loop systems. They show that the average path loss characteristic can be separated into two regions, which gives rise to a two slope model. They show that the diffraction is well approximated by a log-normal distribution.

The second section of this book, Antennas, consists of six papers. The first paper, *Active Microstrip Antenna for Personal Communication System* by M. Wnuk, W. Kolosowski, M. Amanowicz, and T. Semeniuk, describes the development of a microstrip antenna having a radiation pattern which limits the electromagnetic field emitted towards a user's head. The second paper, *Co-located, Dual-band, Multi-function Antenna System for the GloMo Universal Modular Packaging System* by J. S. McLean, J. A. LaCoss, J. R. Casey, E. Guzman, G. E. Crook, and H. D. Foltz, discusses the packaging system for the ultra-high density handheld data terminal. A multi-function antenna, allowing simultaneous operations of two or more radio systems is described. The system was configured to minimize co-site interference. The third contribution is entitled *Self-Calibration Scheme for Antenna Arrays Using the Combined Array Signal* was written by Mark Wiegmann. The calibration employs a beamforming network and a single receiver. A simulation study showed good performance of the calibration algorithms. H. Novak contributed the fourth paper in this group. His paper, entitled *Switched Beam Adaptive Antenna Demonstrator for UMTS Data Rates*, describes the development of a switched beam adaptive antenna system. His system supports data rates in excess of 1 Mbit/s. The fifth paper in the antenna section is entitled *UMTS Radio Network Simulation with Smart Antennas* and was co-authored by B. O. Adrian and S. Haggman. Their simulation study shows substantial capacity improvements in a DS-CDMA network using smart antenna technology. The sixth and final paper in the group of papers dealing with Antennas is entitled *Methods for Measuring and Optimizing Capacity in CDMA Networks Using Smart Antennas,* and was co-authored by S. D. Gordon, M. J. Feuerstein, and M. A. Zhao. The contribution of this paper presents a technique for estimating the forward link capacity of a CDMA system. Their model slows a 27% improvement in capacity over a conventional antenna system.

The third group of papers presented here deals with Multi-Detection. There are three papers in this section. The first of these, *Adaptive Radio Resource Control via Cascaded Neural Networks for Sequenced Propagation Estimation and Multi-User Detection in Third-Generation Wireless Networks* by W. S. Hortos, makes use of a neural network approach to predict radio propagation characteristics and multi-user interference, and to evaluate their impact on wireless networks. The neural network architecture proposed by the author is used to allocate network resources and optimize quality-of-service. The second paper in this section is by M. Golanbari and G. E. Ford. Their contribution, entitled *Successive Interference Cancellation for Interception of the Forward Channel of Cellular CDMA Communications,* considers successive interference cancellation techniques to simultaneously detect cochannel signals in an IS-95 CDMA system. A host of channel impairments are considered. They show performance that tracks the performance of the optimum receiver. In addition, their receiver is near-far resistant. The third and final contribution dealing with multi-user receivers, co-authored by A. Boariu and R. E. Ziemer, is entitled *A New Muliuser Detector for Synchronous CDMA Systems in AWGN Channels*. Boariu and Ziemer introduce a decorrelating decision-feedback multiuser detector based on Cholesky factorization. Simulation results show that the Cholesky-iterative decoder outperforms the standard decorrelating decision feedback detector.

The fourth group of papers in this book treat a variety of technology issues relating to the implementation of radio systems. There are four papers in this group. The first of these, entitled *Modeling Study to Determine the Realistic Constraints of the Wireless Land Mobile Radio Narrowband CAI Interface Specified in the TIA-102 Standard,* is contributed by S. E. Bartlett and K. M. Syed. They describe the result of a channel performance study focusing on the interoperability of the common air interface of the TAI-102 narrowband standard for public safety land mobile radios. Then next paper is by N. L. Marran and is entitled *Over-the-Air Subscriber Device Management Using CDMA Data and WAP*. This contribution illustrates how wireless service providers and their customers can benefit by the deployment of OTA services.

The third paper in this group is entitled *Hyperactive Chipmunk Radio* and was co-authored by G. H. McGibney and S. T. Nichols. The chipmunk radio modulates voice signals in a manner that causes radio waves to behave in the medium as sound waves behave in an acoustic medium. The result is that radio signals inherit many of the desirable characteristics of acoustic voice signals including resistance to both flat and frequency selective fading. The final paper in this section, *Turbo Code Implementations on Fixed Point DSP's*, by E. Cress and W. J. Ebel, considers the implementation of turbo decoding algorithms on the TMS3206201 fixed point DSP architectures.

The fifth group of papers presented in this book deals with wireless data systems. There are three papers in this group. The first of these, *TCP with Adaptive Radio Link*, by D. Huang and J. J. Shi, treats the performance of circuit-switch based TCP over a wireless link. They propose an adaptive radio link protocol to maintain TCP performance under a variety of channel conditions. The following paper, *Reducing Location Update and Paging Cost in a PCS Network* by P. G. Escalle, V. C. Giner and J. M. Oltra, deals with mobility tracking strategies. They propose a new technique that is a hybrid between global and local strategies. The last paper in this section, *Performance Enhancement for TCP/IP on Wireless Links*, by J. S. Stadler, J. Gelman, and J. Howard, discuss the reasons for reduced levels of TCP/IP performance, and describes two techniques for improving performance. Both of the new techniques have been prototyped and tested and both show nearly optimal performance.

The final section of this book contains two invited posters. The first of these, *Development and Implementation of an Adaptive Error Correction Coding Scheme for a Full Duplex Communications Channel,* was co-authored by J. W. Waterston, S. Wooten, W. Bennett, and T. B. Welch. They consider an adaptive coding strategy in which the rate of a $n = 63$ BCH code is adjusted according to channel conditions. They show an increased throughput for a slowly fading Rayleigh channel. The second paper in this set of papers, *Simulink Simulation of a Direct Sequence Spread Spectrum Differential Phase Shift Keying SAW Correlator* by S. M. Nabritt, M. Qahwash, and M. A. Belkerdid, considers the use of a SAW-based demodulator for direct sequence spread spectrum signals. Simulation results agreed well with results obtained from the hardware implementation.

A successful symposium, and consequently the papers contained herein, result from the significant efforts of a dedicated team of people. First, thanks go to those who submitted papers and attended the symposium. Without a strong technical program, the symposium could not continue to prosper. We also thank the MPRG support staff and graduate students. The efforts of Jenny Frank, who took the lead in organizing the symposium and tending to the vast quantity of details associated with the symposium, are gratefully appreciated.

We also acknowledge the support of our technical co-sponsors. These include the IEEE Communications Society, the IEEE Virginia Mountain Section, and the Virginia Tech Student Joint-Chapter of the IEEE Communications and Vehicular Technology Societies.

Blacksburg, Virginia

William H. Tranter
Brian D. Woerner
Jeffrey H. Reed
Theodore S. Rappaport

Very Near Ground RF Propagation Measurements and Analysis

Thad B. Welch[†], Michael J. Walker[‡], and Ray A. Foran[†]

[†]United States Naval Academy
Department of Electrical Engineering {MS-14B}
105 Maryland Avenue, Annapolis, MD 21402-2005
t.b.welch@ieee.org

[‡]United States Air Force Academy
Department of Electrical Engineering
USAFA, CO

Abstract - We analyze and measure the effects associated with placing a cordless phone antenna, with three different orientations, very near the ground (3 - 28 cm). A significant decrease in signal strength occurs when a user falls from the sitting position to the prone position. As much as a 12 dB decrease in signal strength can occur. This information, if available to an injured cordless phone user, could allow for a successfully completed 911 call.

1. INTRODUCTION

When and where available, the traditional *plain old telephone service* (POTS) provides *almost guaranteed access* to 911 and other emergency services. With an increase in cellular and cordless phone usage, more people are relying on these products for both their routine and emergency communication needs. While cellular and cordless products offer increased mobility, challenges associated with the mobile radio channel prevent them from providing *almost guaranteed access* to emergency services. A number of indoor and outdoor emergency scenarios can be proposed. While others have investigated the issue of carrier frequency selection for communication systems with a low antenna height, e.g. [1], we will investigate the effects of antenna height and orientation on system performance. Specifically, we will investigate the scenario of an incapacitated person lying on the floor or ground. If we assume that this scenario exists and that an individual is lying on their back with access to a cellular or cordless phone, then a single antenna phone system would have the tip of its antenna very near the ground plane (floor or ground). Depending on the physical construction of the phone, the antenna could be vertically, horizontally, or diagonally (inverted) oriented relative to the ground plane. The proximity of the antenna to the ground plane suggests that a significant performance degradation may exist [2]. Indeed, it is already known that a *dipole's* impedance fluctuates with varying height above the ground plane. Additionally, this effect is more pronounced if the antenna has a horizontal orientation.

We will consider an indoor scenario where the fixed base station is communicating with a cordless phone which is either in the same room, in an adjacent hallway, or in a distant hallway. At each of the locations, the signal strength will be measured for both the sitting and prone system user. Antenna orientation will also vary from vertical to horizontal, and to diagonal (inverted). Data gathered at the three locations will allow for a comparison of system performances with user elevation and antenna orientation as the only variables. An analysis of the antenna pattern and impedance will also be conducted to help explain the reception difficulties.

We will consider the geometries shown in Figure 1. In Figure 1, a fixed base station labeled "trans" communicating with a system user who is either inside the same room (labeled "pt. 1"), just outside the room in a hallway (labeled "pt. 2"), or further down the same hallway (labeled "pt. 3"). The base station will always remain in the same position. This will place the transmitting antenna's tip 1 meter above the ground.

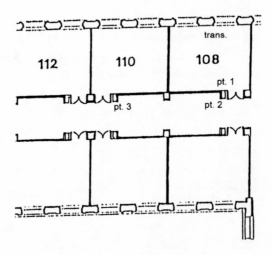

Fig. 1. Measurement geometry.

The system user will be sitting on the floor or in a prone position. Antenna orientation, while the system user is sitting on the floor, will always be vertical. Antenna orientation while the system user is prone will be either vertical, horizontal, or diagonal (inverted). This will place the receiving antenna's tip, 28, 15, or 3 centimeters above the ground, respectively (11, 15, or 16 centimeters above the ground for the antenna's feed point). The fixture that holds the receiving antenna was designed to model a hand-held cordless phone being held to the system user's ear.

2. ANALYSIS

An analysis of the radiating antenna can explain some of the effects seen in the measurements below. The ground can be modeled as an infinite planar boundary. This is a reasonable assumption because the antenna heights and radiation distances are so small compared to the radius of the earth and the measurement sites were essentially flat in the immediate area [3]. When an antenna radiates in the presence of an infinite, planar boundary, some of the energy will propagate directly to the receiver and some will reflect off of the boundary to the receiver. The reflected energy can be modeled as if it is coming from an image source located at the same distance below the boundary as the height of the actual antenna above the boundary, but propagating through free space the entire distance. In the case of a perfectly conducting boundary, all of the energy is reflected and the magnitude of the image will be identical to the source. When the antenna is polarized horizontally, there will also be a $180°$ phase shift. The reflection coefficient, the ratio of reflected energy to incident energy, is constant and equal to either $+1$ or -1. The only effect of the perfectly conducting boundary on the total antenna pattern is the multiplication of an array factor term corresponding to a two-element array with a separation of twice the original source's height.

$$R_V = \frac{\eta_1 \cos\theta_i - \eta_2 \cos\theta_t}{\eta_1 \cos\theta_i + \eta_2 \cos\theta_t}$$

$$R_H = \frac{\eta_1 \cos\theta_t - \eta_2 \cos\theta_i}{\eta_1 \cos\theta_t + \eta_2 \cos\theta_i}$$

$$\eta_1 = \sqrt{\frac{\mu_0}{\varepsilon_0}}$$

$$\eta_2 = \sqrt{\frac{j\omega\mu_0}{\sigma + j\omega\varepsilon_0\varepsilon_r}}$$

$$j\omega\sqrt{\mu_0\varepsilon_0}\sin\theta_i = \sqrt{j\omega\mu_0(\sigma + j\omega\varepsilon_0\varepsilon_r)}\sin\theta_t$$

4

When the medium below the boundary has a finite conductivity, as the ground actually does, the reflected energy can still be modeled as being radiated from an image source, but the net effect changes in several ways. For a finite conductive surface, the reflection coefficient will be complex. The magnitude will almost always be less than one and there will be an additional phase component added. Both the magnitude and phase of the reflection coefficient will also be a function of angle, frequency and polarization.

The effect on the total antenna pattern is the multiplication of a term that accounts for the difference in distances between the source and the receiver and the image and the receiver as well as the image's magnitude and phase.

The power pattern resulting from the summation of the original and image source's radiation in the vertical case is shown below, Fig. 2. The effect of the antenna element's own pattern is not included. Both data sets were normalized to the maximum value of the perfect conducting case. Thus, the power pattern for the finite conducting case is reduced in two ways; because it will never reach the same maximum and because of the altered pattern shape. From this plot we predict a 5-10 dB power reduction due to the loss effects of the finite conductivity of the ground in the angular region of interest.

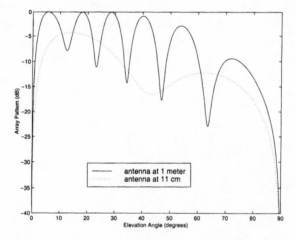

Fig. 2. Array pattern for the transmitter at 1 meter elevation, 10 meter separation, over a reinforced concrete slab (vertical antenna orientations).

The power pattern resulting from the summation of the original and image source's radiation in the horizontal case is shown below, Fig. 3. Notice that there is an additional dependence on the azimuthal direction to the receiver. Again, the effect of the antenna element's own pattern is not included and all data is normalized to the maximum value of the perfectly conducting case. From this plot we predict a negligible reduction, if not a small gain, in the power pattern for the horizontal case.

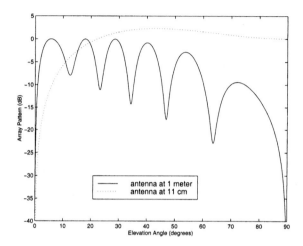

Fig. 3. Array pattern for the transmitter at 1 meter elevation, 10 meter separation, over a reinforced concrete slab (vertical antenna orientation - transmitter and receiver (1 meter elevation), horizontal antenna orientation - receiver (11 centimeters elevation)).

There is another effect that was not thoroughly analyzed but deserves to be mentioned. The presence of an infinite, planar boundary beneath a radiating antenna also alters the antenna's input impedance. If an antenna is connected to a system that is tuned to deliver maximum power based on the antenna's impedance in free space, this change will cause a mismatch and reduce the total radiated power. Using data calculated for dipoles, the effect of this mismatch on vertically oriented antennas is very small. However, the effect on horizontally oriented antennas can be losses on the order of 5-10 dB [4]. We hope to analyze this effect more rigorously in future efforts.

3. DATA GATHERING AND DATA REDUCTION

At each of the three data gather points the receiving antenna fixture was moved approximately 20 wavelengths. The 20 wavelength measurement track was used to be consistent with the results in [5]. During this motorized movement of the antenna, the spectrum analyzer gathered signal strength data and recorded this data, via the HPIB, to the attached laptop PC. At each of the three points, four data sets were gathered. These four sets correspond to the four elevation and antenna orientations combinations of concern (sitting on the floor with a vertical antenna, prone with a vertical antenna, prone with a horizontal antenna, and prone with a diagonal (inverted) antenna). At each of the points 2000 signal strength measurements were gathered into a data set.

Using the cumulative distribution function (CDF) technique discussed in [6], the Rician k factor for the data sets can be calculated. Tables 1, 2, and 3, provide the average path loss, path loss standard deviation, an estimate of the Rician k factor, and the mean-squared error (mse) associated with this best fit for these three data points. We are using a mse of less than 0.0005 to indicate an extremely good fit [7].

At point 1 we can see a 2.4 to 4.6 dB decrease in average signal strength as the system user falls from a sitting position to a prone position. At point 2 we can see a 0.9 to 3.4 dB decrease in average signal strength as the system user falls from a sitting position to a prone position. At point 3 we can see a 3.3 to 4.6 dB decrease in average signal strength as the system user falls from a sitting position to a prone position.

data point 1	average path loss	path loss std dev	Rician k factor	mse
sitting vertical	-12.2 dB	5.8 dB	0.0	0.0000896
prone vertical	-16.8 dB	6.4 dB	0.0	0.000696
prone horizontal	-14.6 dB	5.3 dB	0.0	0.000577
prone diagonal (inverted)	-15.2 dB	5.8 dB	0.0	0.000111

Table 1. Point 1 data analysis.

data point 2	average path loss	path loss std dev	Rician k factor	mse
sitting vertical	-16.2 dB	5.3 dB	1.2	0.000237
prone vertical	-19.6 dB	5.7 dB	1.1	0.000356
prone horizontal	-17.1 dB	5.2 dB	1.2	0.000518
prone diagonal (inverted)	-18.8 dB	5.7 dB	0.8	0.0000802

Table 2. Point 2 data analysis.

data point 3	average path loss	path loss std dev	Rician k factor	mse
sitting vertical	-20.2 dB	5.1 dB	1.4	0.000305
prone vertical	-24.6 dB	4.8 dB	2.0	0.000312
prone horizontal	-23.5 dB	6.0 dB	0.0	0.000409
prone diagonal (inverted)	-24.8 dB	4.8 dB	1.1	0.000312

Table 3. Point 3 data analysis.

All of this data was taken in Maury Hall on the United States Naval Academy. This building houses the Department of Electrical Engineering and is very convenient for this type of data collection effort. We were very concerned about the structural differences between this building and a traditional home, since Maury Hall was build in the early 1900's with exterior walls made of stone and brick totaling over 3 feet thick and with floors that are almost two feet thick made of concrete and brick. Despite the massive structural difference, similarities can be found with modern construction techniques on the interior partitioning walls. Maury was last renovated in the mid-1970's, during which partitioning walls were replaced by sheet rock over steel stud construction. A comparison between the data taken in Maury Hall and a data taken in a more traditional home was still needed. To allow for this comparison, similar measurements were taken in a private residence. Point 4 was the first of the two residential measurements taken. This point was in the same room as the transmitter, 3.6 meters away. Point 5 was the second of the two residential measurements taken. This point was in a nearby room, 10 meters away. Tables 4 and 5, provide the average path loss, path loss standard deviation, an estimate of the Rician k factor, and the mean-squared error (mse) associated with the best fit for these two data points.

data point 4	average path loss	path loss std dev	Rician k factor	mse
sitting vertical	-6.6 dB	2.6 dB	3.5	0.000160
prone vertical	-13.7 dB	5.7 dB	1.2	0.000406
prone horizontal	-13.5 dB	4.9 dB	1.6	0.0000808
prone diagonal (inverted)	-10.2 dB	4.0 dB	2.0	0.000249

Table 4. Point 4 data analysis.

data point 5	average path loss	path loss std dev	Rician k factor	mse
sitting vertical	-27.2 dB	5.1 dB	1.2	0.000337
prone vertical	-39.1 dB	5.1 dB	0.0	0.000670
prone horizontal	-32.5 dB	4.1 dB	2.4	0.000146
prone diagonal (inverted)	-37.4 dB	6.0 dB	0.0	0.00102

Table 5. Point 5 data analysis.

At point 4 we can see a 3.6 to 7.1 dB decrease in average signal strength as the system user falls from a sitting position to a prone position. Unlike the Maury Hall point 1 measurement results, the Rician k factor at point 4 indicated a significant specular component within the propagation. At point 5 we can see a 5.5 to 12.1 dB decrease in average signal strength as the system user falls from a sitting position to a prone position. This represents the largest decrease in average signal strength at any of the 5 data points.

4. CONCLUSIONS

The finite conductivity of the ground cannot be ignored in the near-ground radiation problem. The lossy ground directly generates two important effects, antenna mismatch and a complex reflection coefficient, that together roughly approximate the measured power losses. These and other effects lead to a significant decrease in the signal strength when a cordless phone user falls from a sitting position to the prone position. This type of analysis and propagation information will be useful to the designers, manufacturers, and end users of cordless, cellular, and PCS phones. Should an emergency arise that requires the prone user of such a system to place a 911 call, these same or similar geometries will exist. User knowledge of the fact that path loss generally decreases with the user's antenna elevation could allow for a successfully completed 911 call.

The array pattern prediction provided is more appropriate for an outdoor scenario than for the residential scenario we measured.

This type of analysis and propagation information can also be used by military ground units that are required to stay in contact with base stations while not revealing their position. Extensions to other geometries and frequencies will also allow similar evaluations for cellular and PCS systems.

REFERENCES

[1] R.F. Graham, Jr., "Identification Of Suitable Carrier Frequency For Mobile Terrestrial Communication Systems With Low Antenna Heights," *Proc. MILCOM'98*, pp. 1-5 of session 9.3 [CD-ROM], Oct. 1998.

[2] W.C. Jakes (editor), *Microwave Mobile Communication*, IEEE Press, New Jersey, 1994 (originally printed in 1974).

[3] R.E. Collins and F.J. Zucker (editors), *Antenna Theory - part 2*, McGraw-Hill, New York, 1969.

[4] C.A. Ballanis, *Antenna Theory - Analysis and Design*, John Wiley & Sons, New York, 1997.

[5] T.S. Rappaport, *Wireless Communications, Principles and Practices*, Prentice Hall PTR, New Jersey, 1996.

[6] J.D. Parsons, *The Mobile Radio Propagation Channel*, John Wiley & Sons, Inc., New York, 1992.

[7] R. Kattenbach and T. Englert, "Investigation Of Short Term Statistical Distributions For Path Amplitudes And Phases In Indoor Environments," *Proc. VTC'98*, pp. 2114-2118, session 64-4 [CD-ROM], May 1998.

Identification of Time-Variant Directional Mobile Radio Channels

R.S. Thomä, D. Hampicke, A. Richter, G. Sommerkorn, A. Schneider, U. Trautwein

Department of Electrical Engineering and Information Technology
Ilmenau University of Technology
P.O.B. 100565, D-98684 Ilmenau, Germany

Phone: (+49) 3677-692622, Fax: (+49) 3677-691113, E-mail: tho@e-technik.tu-ilmenau.de
http://www-emt.tu-ilmenau.de

Abstract

For the real-time identification of the time-variant, directional structure of the mobile radio channel impulse response, a broadband vector channel sounder is described. The measurement procedure relies on periodic multi-frequency excitation signals, correlation processing and joint delay-azimuth superresolution based on the ESPRIT algorithm. The underlying signal model is developed and the different possibilities of ESPRIT application are discussed. Problems of imperfect receiver and antenna performance and the appearing resolution limits are outlined. Results of multidimensional correlation analysis of various channel scenarios in the Doppler-delay-angular domain are presented.

1 Motivation

Efficient wireless transmission constitutes a key technology for future universal mobile communication systems. High data rates, adequately defined quality of service, high system capacity and high bandwidth efficiency require new and sophisticated radio link designs. That includes adaptive equalization and adaptive modulation schemes. Most recently, smart antenna principles are considered to enhance system performance. The expected benefits include increased capacity and quality of service as a result of interference reduction by spatial filtering and sophisticated equalization and diversity procedures in the joint delay and angular domain [1].

Design and simulation of smart antenna modems requires profound knowledge of the radio channel impulse response (CIR) statistics [2]. The multipath components of the time-variant impulse response have to be analyzed with respect to their path delays and directions of arrival. Wideband, real-time measurement of the time-variant directional radio channel is a very demanding task. High multipath time delay and angular resolution as well as fast measurement repetition rate are required. Traditional measurement methods based on sweeped network analyzers or sliding correlators and rotating antennas are generally not suited since they presume time-invariant radio channels.

The paper describes the basic performance of the antenna array based Vector Radio Channel Sounder RUSK ATM. The parametric signal model and the estimation procedure are outlined. Finally, an introduction to the statistical analysis of the channel characteristics and some measured examples are given. The channel sounder has been developed under the german national project line ATMmobil which is designated for next generation broadband multimedia mobile radio systems.

2 Signal model

The signal transmission in a typical mobile radio channel is affected by time-angular-variant multipath propagation as indicated in Fig. 1. In the uplink the waves impinging on the base station (BS) antenna consist of a line-of-sight component (LOS) and contributions from K-1 non-LOS paths with relative excess delays from different directions that result from scattering, reflection or diffraction. The individual path contributions are time-varying because of mobile station (MS) and environmental objects movement. Generally, the path weights $\gamma_k(t)$ may be fast fading since in some microscopic sense (not resolved by the respective measurement resolution in time and/or azimuth) any scattered and diffracted path consists of a superposition of time-variant contributions. Therefore, for some limited observation time, the channel impulse response is often considered as a wide-sense stationary stochastic process (WSS). For a longer observation time and MS traveling distances of much more than (typically) some tens of the carrier wavelengths, the mean path time-delays τ_k (TDOA), the directions of arrival θ_k' (DOA), the Doppler shifts α_k and the dominant path numbers K are varying and the r.m.s. path weights become slowly fading because of the changing scenario geometry and possible path shadowing and varying path loss. Thus, the time-space-variant impulse response can be given as:

$$h_{t,s}(t,\tau,s) = \sum_{k=1}^{K} \gamma_k(t) e^{-j2\pi t \alpha_k}\, e^{-j2\pi s\theta_k}\, \delta(\tau - \tau_k) \tag{1}$$

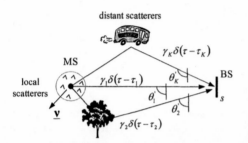

	short-term variant	long-term variant
γ_n : path weight	✓	✓
α_k : Doppler shift		✓
τ_k : time-delay		✓
θ_k' : direction of arrival		✓
K : number of multipath components		✓

Fig. 1: Time-angular-variant multipath propagation in a mobile radio link.

The channel response in (1) represents the channel in the equivalent baseband domain. The parameters have to be considered as valid in some limited frequency range only. Since we restrict our discussion to azimuthal DOA, the space domain variable s is appropriately defined by the linear BS antenna aperture. We also assume plane wavefronts. E.g., for object distances $r > 28.5\ S\lambda_0$ we get up to $\pm 0.5°$ phase error along the aperture [3]. Furthermore, the wavefront delays along s can be approximated by the complex phasor multiplication $\exp(-j2\pi s\theta_k)$ if the signals are narrowband and the antenna aperture is small enough, $S\lambda_0 \ll c/B$. Here c is the velocity of light, B is the measurement bandwidth and S is the maximum antenna aperture. Both, s and S are given normalized to the wavelength λ_0 at the carrier frequency. The projection of the waves from azimuthal DOAs θ' to the antenna aperture results in the directional cosines $\theta_k = \cos(\theta_k')$.

The parametric linear input/output signal model from (1) is more clearly arranged if we use the appropriate Fourier transform relations

$$\alpha,\tau,\theta \quad h(\alpha,\tau,\theta) = \sum_{k=1}^{K} \Gamma_k(\alpha) * \delta(\alpha - \alpha_k)\, \delta(\tau - \tau_k)\delta(\theta - \theta_k)$$

$$t,f,s \quad H(t,f,s) = \sum_{k=1}^{K} \gamma_k(t) e^{-j2\pi t \alpha_k} e^{-j2\pi f \tau_k} e^{-j2\pi s\theta_k} \tag{2}$$

The Doppler-azimuth-variant impulse response is related to the time-space-variant frequency response by a 3-D Fourier transform. Note, that it is the linear array assumption that results in a Fourier transform relation for the space-azimuth branch of the transform. The time-Doppler Fourier pair shows that fast fading path weights $\gamma_k(t)$ introduce local Doppler spread. If single reflections are resolved, that results in pure Doppler shift and constant path weight magnitudes. Only then the channel parameter estimation procedure may be considered as a three-dimensional harmonic retrieval problem.

3 Vector radio channel sounder hardware design

Depending on the available hardware, the measurement of the system response functions in (2) can be performed in any domain of the transform pairs. By all means, the resolution of the parameter triple α_k, τ_k, θ_k is given by the resulting aperture sizes T,B,S in the t, f, s-domain which are limited by the signal model restrictions and the hardware constraints. The latter are imposed by the instantaneous bandwidth B from ADC/DAC sampling rate limitations and by the antenna aperture from receiver channel number limitations. At the same time the narrow band modeling assumption required to establish (1) imposes constraints on the bandwidth and the antenna aperture. The maximum measurement time aperture T is limited by the invariance condition of the channel parameters. On the other hand, the measurement repetition rate has to be fast enough in order to reproduce the time variation by meeting the Nyquist sampling criterion w.r.t. variable t which is given by the expected maximum Doppler bandwidth. Therefore, the hardware design for a real-time vector channel sounder is a demanding task that requires very fast processing. The remaining margin for measurement time stretching by cost saving sequential operations is given by the delay-Doppler spread factor, which is defined by the product of the maximum excess delay and the maximum Doppler bandwidth. For a typical mobile radio channel, as a result of the path loss and the specified maximum transmit power and because of the maximum Doppler bandwidth, it is well below 1 %. In the terminology of time-variant linear systems [4], the mobile radio channel is clearly "underspread". Since the vector channel sounder RUSK ATM relies on real-time sampling instead of sliding correlation, sequential operation in the spatial domain can still be used without sacrificing the advantage of having full real-time access to the time-variant radio channel. By fast antenna multiplexing, the individual antenna responses that form the channel response vector snapshot (CRVS) are sequentially estimated. The multiplexer timing is synchronized to consecutive periods of the Tx signal. Since only a single RF downconverter channel is required, the hardware expense is reduced dramatically as compared to a completely parallel multichannel operation. Details of the hardware design are described in [6]. Table 1 gives an overview of the resulting hardware parameters.

Instantaneous bandwidth	120 MHz		
Frequency range	5 ... 6 GHz (extension available)		
Scalar impulse response	Response length: 0.8 ... 25.6 μs Dynamic range: 35 dB		
Antenna array	8 elements, multiplexed		
Measurement mode	Event triggered	Standard Doppler	Fast Doppler
Repetition rate Time record length (8 channels, 0.8 μs impulse response)	50 Hz limited by external DAT capacity	1 kHz ≈ 60 s	≈ 70 kHz 256 snapshots
Tx/Rx sync	Rubidium reference / optical fiber		
Interfaces	DAT streamer, SCSI, TFT-Display, telemetry, GPS / DGPS, odometer and gyrosscopic sensor interface		

Table 1:   Basic hardware parameters of RUSK ATM vector channel sounder.

The measurement setup consists of a mobile transmitter (Tx) that acts as the MS and of a fixed receiver (Rx) that plays the role of the BS. 120 MHz bandwidth periodic multi-frequency excitation signals with minimum crest factor are used. Those signals are very effective for frequency domain system identification since they offer an exactly limited frequency spectrum, allow fast measurements and low estimation variance as well as minimum leakage bias in case of synchronized FFT processing [5]. The signal is generated in the RF-range by upconversion from the baseband and radiated with a power of 26 dBm from an omnidirectional monopole antenna. The 5.2 GHz Rx antenna is formed by an M=8 element uniform linear array (ULA) of $\lambda_0/2$ spaced planar elements, which are vertically polarized and have 120° azimuthal beamwidth. The choice of the antenna array geometry strongly influences the DOA estimation performance. As will be seen later, a ULA geometry allows very effective algorithms from a computationally point of view. For any array output $m = 1 \ldots M$, the receive signal spectrum $\hat{Y}(n, \mu, m)$ is calculated by FFT and an estimate of the CRVS in the frequency domain is calculated as $\hat{H}(n, \mu, m) = \hat{Y}(n, \mu, m) / \hat{X}(n, \mu, m)$ with the argument variables n, μ denoting the snapshot time instant nt_0 and the frequency $\mu f_0 = \mu / t_p$, resp. The excitation signal reference spectrum $\hat{X}(n, \mu, m)$ is measured by a back-to-back calibration procedure where the radio channel is replaced by an attenuator connected between Tx and Rx. Therefore, Tx- and Rx-frequency response and nonlinear distortion of the Tx power amplifier are removed from the measurement results [13].

4 Channel parameter estimation

For evaluation of micro- and especially of picocell-scenarios, very high path parameter resolution is required. Even it aperture sizes in the three domains are chosen as large as possible, simple DFT estimation of the discrete parameters in (2) would not yield satisfactory results. Firstly, angular resolution is limited by the array aperture to $0.89/S$ which corresponds to about 12.5° of DOA resolution in case of the $S = 4$ array at broadside direction. It even reduces to about 30° at the skirts of the beam sector. Secondly, the τ_k resolution in the delay domain is in the order of 8 to 15 ns which corresponds to 2.4 to 4.5 m, depending upon the window function in the frequency domain that is used for CIR sidelobe reduction. Even Doppler resolution can be a problem when the time aperture length is strongly limited, especially for rapidly changing scenarios. To overcome the DFT resolution limits, parametric DOA estimation is applied in order to achieve superresolution. From the different procedures [8], the ESPRIT-type algorithms are especially suited for ULAs. As compared to other algorithms, such as Maximum Likelihood (see e.g. [9] for an effective iterative implementation for channel sounder application), the ESPRIT algorithm (Estimation of Signal Parameters via Rotational Invariance Techniques) is very time-effective since it avoids extensive multidimensional search. The matrix model for estimation is given as:

$$\mathbf{H}(n, \mu) = \mathbf{A}(\theta) \mathbf{\Gamma}(n, \mu) + \mathbf{N}(n, \mu) \tag{3}$$

with the frequency domain CRVS vector $\mathbf{H}(n, \mu) = [\hat{H}(n, \mu, 1) \ldots \hat{H}(n, \mu, M)]$, the spatially white noise vector $\mathbf{N}(n, \mu)$ and the impinging wavefront vector $\mathbf{\Gamma}(n, \mu) = [\gamma_1 \exp(-j2\pi \mu f_0 \tau_1) \ldots \gamma_K \exp(-j2\pi \mu f_0 \tau_K)]^T$. The $M \times K$ array response matrix $\mathbf{A}(\theta)$ is composed of the response vectors for the K individual wavefronts $\mathbf{a}(\theta_k) = [1, \exp(-j2\pi d\theta_k) \ldots \exp(-j2\pi(M-1)d\theta_k)]^T$ where d gives the distance between the antennas normalized to the wavelength λ_0. (3) represents the discrete, measured representation of the signal model (2) in the t, f, s-domain. Since the parametric approach can be considered as an alternative to the DFT, there are several possibilities to use superresolution algorithms for estimating the model parameters τ_k, θ_k, α_k. E.g., only a single Fourier transform branch can be replaced by a 1-dimensional ESPRIT estimation. Since the most severe resolution restriction seems to be imposed by the antenna aperture, we start with the space (s) to azimuth (θ) transform. Fig. 2 gives an idea of the resulting procedure. With respect to the estimation of the remaining parameters τ_k, α_k the DFT-approach is used. In order to simplify the presentation, only the delay-azimuth estimation is shown and the time-Doppler domain is omitted. The first step is to transform the frequency domain CRVS to the delay domain by DFT. Then a 1-D ESPRIT DOA estimation is applied for all τ_k bins that contain enough energy. This explains how high measurement bandwidth supports resolution of a large number of paths since only those paths that show the same path delay (within the delay resolution limit) have to be resolved in the azimuthal domain. The picture, however, also explains that the τ_k, θ_k estimation task can more consequently be characterized as a 2-D joint delay-angular estimation problem.

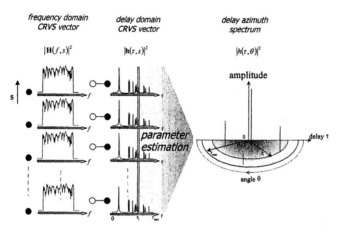

Fig. 2: Delay-azimuth snapshot estimation from one frequency domain CRVS.

Like other DOA estimation algorithms, the ESPRIT belongs to the subspace class that exploits the eigenvector structure of the array covariance matrix in order to estimate the signal subspace. Its efficiency, however, comes from the usage of the special structure of the array response matrix. In case of an ULA, $A(\theta)$ takes the form of a Vandermonde matrix. The main idea of the ESPRIT is to divide it into two submatrices that correspond to identical sub-arrays which may even be arranged partly overlapping in space. Then there exists a projection matrix that uniquely rotates the output of one sub-array to the other. The actual estimation problem is now reduced to find that projection matrix by solving a general least-squares problem (LS-ESPRIT) of the resulting (typically) overdetermined set of equations. The eigenvalues of that projection matrix are directly related to the DOAs. Estimation accuracy can be enhanced by using structured or total last squares (SLS-ESPRIT, TLS-ESPRIT). The standard ESPRIT approach is even outperformed by the recently introduced unitary ESPRIT algorithm [10], [11]. That procedure transforms the subspace estimation step to a real problem by exploiting the structure of centro-symmetric arrays (as it is given by an ULA). Thereby, the estimated twiddle factors are constraint to the unit circle which reduces estimation errors. Moreover, unitary ESPRIT may be arranged to inherently contain forward-backward averaging. A further advantage of that idea is that it can be extended to a closed form 2-D joint parameter estimation algorithm that provides automatically paired sets of parameters. That allows a very smart solution of the joint delay-azimuth estimation problem described by Fig. 2 and (2) whereby also the delay resolution is enhanced by the superresolution capability of the ESPRIT. For more details see [3], [11].

As a result of the underlying stochastic model, proper estimation of the signal subspace is an issue. Since the CRVS (2) is assumed to be WSS with uncorrelated path scattering processes $\gamma_k(t)$ some time domain averaging is required in order to get a stable estimate. The maximum allowable time interval for averaging is given by the invariance condition of the slowly varying model parameters (s. Fig.1). Alternatively to the covariance averaging approach, there exists the direct data approach that uses singular value decomposition (SVD) of the same time sequence interval of the CRVS. That is often preferred because of a better numerical stability. Finally, the ESPRIT ends with an estimate of one set of the mean channel parameters τ_k, θ_k for the time interval considered. Then it is still necessary to estimate the path weights $\gamma_k(t)$. Since they are generally fast fading, they have to be determined consecutively for any single snapshot in time by least squares estimation of K sets of nullsteering beamformer weights using a Penrose-Moore pseudoinverse. The result is a sequence of snapshots in time that directly corresponds to the time-azimuth-variant impulse response $h_t(t,\tau,\theta)$. A Fourier transform w.r.t. time t results in the Doppler-azimuth-variant impulse response of (2).

Unfortunately, the azimuthal signal subspace decomposition fails if impinging wavefront signals are correlated. Since all paths are launched from the same source, reflected signals have to be considered as coherent if they are subjected to nearly the same delay (within the time resolution limits) and if the related scattering process is at least partly deterministic. In particular, the ability to resolve closely spaced paths is reduced dramatically [8]. Then spatial smoothing of the estimated signals subspace has to be performed for a rank enhancement. Since for that purpose the array has to be divided into overlapping subarrays to be smoothed, the effective array aperture is reduced. Thus, the maximum number of sources that can be resolved by the $M=8$ ULA at any delay bin is reduced to about 5. In case of joint delay-azimuth estimation also frequency domain smoothing is required in order to enhance delay subspace separation of paths from (nearly) the same azimuthal DOA. Of course, subspace smoothing not only enhances the rank of the signal subspace, it also improves statistical stability by noise reduction.

The quality of the whole procedure is also strongly influenced by the correct choice of the model order, but that cannot be discussed here.

5 Measurement Errors, Calibration and Resolution

In any practical measurement setup the acquired data are somewhat impaired by limited accuracy, noise and interference. Since a superresolution procedure can be understood as an extrapolation in the corresponding aperture domain, it is very sensitive to measurement errors. Therefore, the achievable resolution is limited by noise and device parameter impairments. As a general rule, amplitude and phase uniformity of the array channels determines the achievable DOA resolution while frequency domain invariability determines the TDOA resolution.

Since the RUSK ATM receiver consists only of a single down-convertor channel, there are strongly reduced problems with unequal receiver channels. Mainly phase noise of the mixer frequency sources is an issue since the antenna outputs are sampled sequentially in time. In the device described phase noise is kept low enough by sophisticated PLL/VCO design. Antenna impairments, however, cause more problems. Because of the close spacing between neighboring elements, parasitic electromagnetic coupling cannot be avoided. That results in severe distortions of the antenna beam patterns. Although ESPRIT does not require the precise knowledge of the array response vector, it relies on identical beam patterns. Any non-uniformity of the beams disturbs the ESPRIT since the algorithm interprets that distortion as a result of impinging waves. Simulation has shown that the maximum peak-to-peak ripple of the resulting beam patterns should be below 0.5 dB in order to achieve 5° angular resolution of coherent paths. That can only be reached with sophisticated antenna array calibration. The calibration procedure is based on a set of reference measurements using a single source under well-defined propagation conditions in an anechoic chamber with constant delay τ_k at an equidistant grid of well known azimuth angles θ_k. Details of an effective eigenvector-based calibration matrix estimation procedure and measured results are given in [12]. Stable 5° resolution of two sources impinging with the same TDOA has been demonstrated over the complete 120° azimuthal antenna sector with some degradation only at the skirts of the beams. Also the most complicated 5 coherent source scenario can be resolved [6]. It has been shown that parasitic echoes during calibration have to be at least 30 dB down if they are not clearly resolved in delay. Otherwise the calibration result is severely impaired.

Imperfect frequency response uniformity of the calibrated device results from remaining internal reflections that may be introduced by mismatch between the calibration and the measurement setup and from slightly changing frequency response as a result of AGC switching. Currently, about 1.5 ns TDOA resolution of sources impinging from the same DOA is reliably achieved which corresponds to about 50 cm spatial resolution.

6 Second Order Statistical Analysis

Because of the underlying stochastic signal model, statistical analysis based on second order correlation can reveal interesting channel features. The WSSUS channel model helps to define a 3-dimensional correlation function by assuming stationarity w.r.t. time, frequency and spatial distance ($\Delta t, \Delta f, \Delta s$). That corresponds to uncorrelated behavior w.r.t. the variables Doppler shift, delay and azimuth (α, τ, θ). The following 3D-Fourier-Transform relates the expected Doppler-delay-azimuth spectrum to the corresponding expected time-frequency-spatial correlation:

$$\begin{aligned} \alpha,\tau,\theta \quad & r(\alpha,\tau,\theta) = \mathrm{E}\left\{ \left| h(\alpha,\tau,\theta) \right|^2 \right\} \\ \Delta t,\Delta f,\Delta s \quad & R(\Delta t, \Delta f, \Delta s) = \mathrm{E}\left\{ H(t,f,s)\, H^*(t+\Delta t, f+\Delta f, s+\Delta s) \right\} \end{aligned} \qquad (4)$$

The estimation of (5) should be based on the spectral domain α, τ, θ since that avoids time-expensive correlogram calculation. From classical spectral estimation of stochastic processes [14] it is well-known that some statistical averaging or smoothing is required in order get stable estimates. The relative variance of the estimation is inversely proportional to some aperture-resolution product. That means, there is a tradeoff of statistical stability against resolution that further aggravates the resolution constraints. From a computational point of view, one possibility is to smooth the rough estimate of the magnitude squared Doppler-azimuth variant impulse response by the smoothing window $W(\cdot)$:

$$\hat{r}(\alpha,\tau,\theta) = \iiint W(\alpha-\alpha',\tau-\tau',\theta-\theta') \left| h_{TBS}(\alpha',\tau',\theta') \right|^2 \mathrm{d}\alpha' \mathrm{d}\tau' \mathrm{d}\theta' \qquad (5)$$

The index in h_{TBS} denotes a *TBS* aperture limited estimate of the Doppler-azimuth variant impulse response (2). The smoothing impact of $W(\cdot)$ is determined by its spread in the different domains which shows the compromise between variance reduction and resolution bias. Therefore, the support of $W(\cdot)$ in each of the different domains has to be chosen deliberately in order to match the desired resolution of the path clusters. If the channel has to be evaluated from the viewpoint of some prospective application system, one objective may be to meet the resolution limits of that system. Another estimation possibility is to divide the time sequence $h_t(t,\tau,\theta)$ with the total aperture record length T into smaller, weighted and overlapping segments and proceed with spectral averaging. In that case, Doppler and delay resolution have to be chosen in advance, but eventually both procedures can be effectively combined [14].

As discussed in section 5, the maximum achievable delay-azimuth resolution corresponds to scattering clusters of about 50 cm in diameter. Therefore, with the carrier frequency of 5.2 GHz, only about

18

10 Doppler cycles are available. In that case it seems not appropriate to stake on DFT Doppler resolution. Then some parametric spectral resolution procedure should be applied to achieve reliable, high resolution estimates from the short time segments. Standard AR estimators seem to be well suited since at the same time a parametric channel model arises that is well suited for statistical channel modeling [17].

The general WSSUS relation (4) offers a large variety of deduced functions by applying Fourier transform relations w.r.t. the different variables as given in Fig 3. Reduced domain functions and parameters are calculated by integration over one, two or three variables such as the Doppler-delay spectrum, the delay-azimuth spectrum, the Doppler-azimuth spectrum, the delay spectrum or the Doppler spectrum, the Doppler spread or angular spread etc.

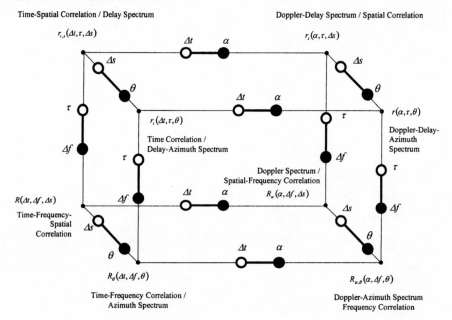

Fig. 3: WSSUS spectral and correlation relations.

7 Measurement examples

In the following, measurement results that are typical for an industrial environment are demonstrated. The measurement campaign took place in a car factory hall of the Daimler-Chrysler AG in Sindelfingen (Germany). Fig. 4, at first, shows the spatially averaged, magnitude-squared CIR sequence that was recorded with the Tx moving away from the Rx between two car assembly lines and subsequently driving around one of them and moving back toward the Rx. The sequence clearly shows the adequately changing delays. It also indicates that the LOS is lost during the way back. It can be expected that the channel parameters will change significantly at that instant since transmission will be based only on scattering and multiple reflection if LOS is obstructed. Fig. 5 shows more details from two cut-outs of the same impulse response at LOS conditions (left) and non-LOS conditions (right), respectively. From Fig. 6 the delay-

Doppler spectra at a LOS and a non-LOS location can be seen. The boundary projections in the 3-D pictures show the max-hold average delay spectrum and Doppler spectrum, respectively, that can be identified by the axis variables. In the non-LOS case an almost ideal classical Jakes Doppler spectrum occurs. Fig. 7a-c displays the short-time averaged delay-azimuth spectrum at two LOS-locations and at one non-LOS location. Here the boundary projections are the max-hold average azimuth spectrum and again the delay spectrum. It can be seen that the angular spread gradually decreases with increasing distance between Tx and Rx antenna during the LOS situation and suddenly increases when LOS is disappearing. The same characteristic is indicated in Fig. 8 where the r.m.s. angular spread and the r.m.s. delay spread are shown for the complete record length.

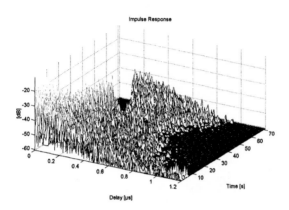

Fig. 4: Spatially averaged, log. magnitude squared impulse response (complete measurement drive).

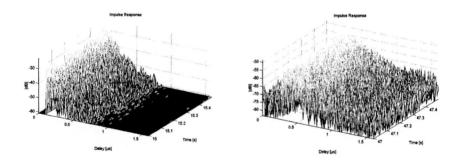

Fig. 5: Spatially averaged, log. magnitude squared impulse response: cut-out at 15 s (left) and at 47 s (right)

20

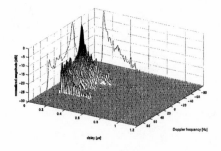
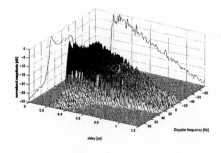

Fig. 6: Spatially averaged Delay-Doppler spectrum at 11 s (left) and at 60 s (right).

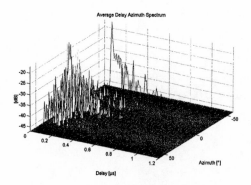

Fig. 7a: Local time averaged delay-azimuth spectrum at 7 s.

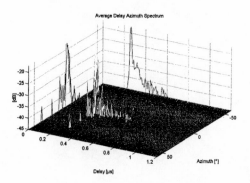

Fig. 7b: Local time averaged delay-azimuth spectrum at 38 s.

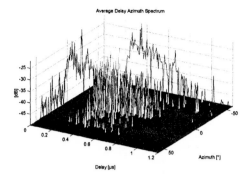

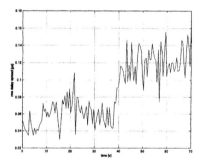
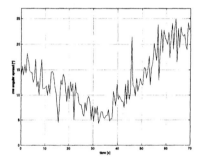

Fig. 7c: Local time averaged delay-azimuth spectrum at 55 s.

Fig. 8: r.m.s. delay spread (left) and r.m.s. angular spread (right) along the complete record.

8 Conclusions

A hardware-effective realization of a real-time vector channel sounder based on fast antenna multiplexing has been shown. That device allows full statistical analysis of the Doppler-delay-azimuth statistic of mobile radio channels. Further investigations will include elevation and polarization analysis as well. Also correlation of the delay-azimuth statistics between different frequency bands is of interest for investigation of uplink- and downlink-beamforming in frequency duplex systems. Estimation of parametric channel models and the usage of the measured channel responses for realistic link level simulation in different scenarios including dynamic changing situations are a further issue [15], [16].

Analysis of the radio channel under different model scenarios needs careful planning, extensive measurement campaigns and intensive analysis of the recorded data [18]. Effective software tools are required for that purpose.

22

Acknowledgements

This work is partly supported by the German Federal Ministry of Education, Science Research and Technology under the ATMmobil project line and partly by the DFG (Deutsche Forschungsgemeinschaft). The authors are grateful to MEDAV GmbH (http://www.medav.de/) for cooperation in development of the Vector Channel Sounder RUSK ATM and to the consortium of the EU project METAMORP for cooperation in channel sounder calibration and channel parameter definition.

References

[1] A.J. Paulraj, C.B. Papadias, "Space-Time Processing for Wireless Communications," *IEEE Sig. Proc. Mag.*, vol.14, no.6, pp. 49-83, Nov. 1997

[2] R.B. Ertel, P. Cardieri, K.W. Sowerby, T.S. Rappaport, J.H. Reed, "Overview of Spatial Channel Models for Antenna Array Communication Systems," *IEEE Personal Comm. Mag.* vol. 5, no. 1, pp. 10-22, Febr. 1998

[3] U. Martin, "Spatio-Temporal Radio Channel Characteristics in Urban Macrocells," *IEE Proc. Radar, Sonar, Navig.*, vol. 145, no. 1, pp. 42-49, Febr. 1998

[4] W. Kozek, "On the Transfer Function Calculus for Underspread LTV Channels," *IEEE Trans. SP*, vol. 45, no.1, Jan. 1997.

[5] R.S. Thomä, H. Groppe, U. Trautwein, J. Sachs, "Statistics of Input Signals for Frequency Domain Identification of Weakly Nonlinear Systems in Communications," *IEEE Instr. and Measurement Technology Conf.*, Brussels, pp. 2-7, Juni 4-6, 1996

[6] R.S. Thomä, D. Hampicke, A. Richter, G. Sommerkorn, A. Schneider, U. Trautwein, "Identification of Time-Variant Directional Mobile Radio Channels," *16th IEEE Instrumentation and Measurement Technology Conference, IMTC/99*, Venice, Italy, May 24-26, 1999, accepted for publication.

[7] K. Schwarz, U. Martin, H.W. Schüßler, "Devices for Propagation Measurement in Mobile Radio Channels," *Proc. of the 4th IEEE Int. Symp. on Personal Indoor and Mobile Radio Communications, PIMRC'93*, Yokohama, Japan, pp. 387-391, Sept. 1993.

[8] H. Krim, M. Viberg, "Two Decades of Array Signal Processing – the Parametric Approach," *IEEE Sig. Proc. Mag.*, vol.13, no.4, pp. 67-94, July 1996.

[9] B.H. Fleury, D. Dahlhaus, R. Hedergott, M. Tschudin, "Wideband Angle of Arrival Estimation Usinf the SAGE Algorithm," *Proc. of the 4th IEEE Int. Symp. on Spread Spectrum Techniques and Applications, ISSSTA'96*, pp. 79-85, Mainz, Sept. 1996

[10] M. Haardt, J.A. Nossek, "Unitary ESPRIT: how to obtain increased estimation accuracy with reduced computational burden," *IEEE Trans. Signal Processing*, vol. 43, pp. 1232-1242, May 1995

[11] M. Haardt, "Efficient One-,Two-, and Multidimensional High-Resolution Array Signal Processing," Shaker Verlag, Aachen, Germany, 1996, ISBN 3-8265-2220-6

[12] K. Pensel, J.A. Nossek, "Uplink and Downlink Calibration of an Antenna Array in a Mobile Communication System," *COST 259 Technical Document, TD(97)55*, Lisbon, Portugal, Sept. 1997

[13] P. H. Lehne (Ed.), "Review of Existing Channel Sounder Measurement Setups an Applied Calibration Methods," Measurement, Testing and Calibration of Advanced Mobile Radio-Channel Equipment (METAMORP), Deliverable META/D-1/TR/D-1/1/b1, June 1998, http://www.nt.tuwien.ac.at/mobile/projects/METAMORP/

[14] S.L. Marple, "Digital Spectral Analysis," Prentice Hall, 1987

[15] U. Trautwein, K. Blau, D. Brückner, F. Herrmann, A. Richter, G. Sommerkorn, R.S. Thomä, "Radio Channel Measurement for Realistic Simulation of Adaptive Antenna Arrays," *The 2nd European Personal Mobile Communications Conference, EPMCC '97*, Bonn, Germany, pp. 491-498, Sep. 30 - Oct. 2, 1997.

[16] U. Trautwein, G. Sommerkorn, R.S. Thomä, "A Simulation Study on Space-Time Equalization for Mobile Broadband Communication in an Industrial Indoor Environment," IEEE Conf. Vehicular Technology, VTC Spring 1999, Houston, Tx, accepted for publication.

[17] U. Martin, "Statistical Mobile Radio Channel Simulator for Multiple-Antenna Reception," IEICE 1996 International Symposium on Antennas and Propagation, Chiba, Japan, pp. 217-220, Sept. 1996

[18] U. Martin, J. Fuhl, I. Gaspard, M. Haardt, A. Kuchar, C. Math, A.F. Molisch, R.S. Thomä, "Model Scenarios for Intelligent Antennas in Cellular Mobile Communication Systems – Scanning the Literature." Submitted to Wireless Personal Communications, Special Issue on Space Division Multiple Access.

Propagation measurements and Simulation for Wireless Communication systems in the ISM Band

B.L. Johnson Jr., P.A. Thomas, D. Leskaroski, and M.A. Belkerdid

Electrical and Comp. Eng. Dept, University of Central Florida, Orlando FL 32816

blj08431@pegasus.cc.ucf.edu

ABSTRACT

Various RF propagation models have been introduced for different frequency bands. These models can be characterized into two different classes. The first class is called deterministic and the second stochastic.

Both of these techniques were utilized in a recent study performed in South Florida. The study consisted of propagation measurements taken in the 2400-2483.5 MHz Instrumentation, Scientific, and Medical (ISM) frequency band. The measurements performed in the study were then put into the Hata-Okumura propagation model. The results from the Hata-Okumura model were then imported into Mathcad. The result of this was a simulation based on propagation model seeded by experimental measurements.

This simulation-based model consists of a propagation prediction model, a terrain database, and a subsystem of various Mathcad programs to simulate coverage patterns in terms of path loss and bit error rate. Simulations are presented based on the prevalent Hata-Okumura propagation model as well as the proposed propagation model.

Overall, results were well within expectations despite propagation measuring constraints. The Mathcad model has been shown to be a viable method of simulation propagation coverage. This approach towards simulation can in the near future provide a speedy and economic service to communications system design engineers.

INTRODUCTION

Propagation models aid in the development of wireless communication networks. A wireless network can be characterized by its basic components. A typical network consists of a transmitter, receiver, and the surrounding environment. Each variable in the network will effect the propagation model that can be used or developed for the given network. A model can be used for a certain frequency band to predict with a high degree of accuracy the nature of surrounding atmosphere.

This work was funded in part by a grant from the Florida Department of Transportation, Contract NO BB-534

There are several models that can extrapolate out to 2.4Ghz band, but we felt that the Hata-Okumura model was the most appropriate for these tests.

Propagation mechanisms such as reflection, scattering, and diffractions always need to be accounted for. This phenomenon is more profound when there is no existing line-of-sight between the transmitting and receiving antennas. Therefore a typical mobile channel is characterized by multipath reception[1].

Predictions of signal strength and propagation coverage area are vital aspects in the design of wireless communications systems. There are three basic approaches utilized in the prediction of signal strength and propagation coverage area. They consist of the empirical approach, statistical approach, and combination of both. The first approach or empirical is the easiest to implement. It requires only the use of theoretical models, however the downside is that the actual terrain is neglected. The second approach, which is the statistical approach, is the most intensive insofar as to the amount of work that is required. It however yields the more accurate results then the empirical approach. The best approach quite possibly could be a combination of both.

There are several models that can be used for this study they are the Hata-Okumura, Walfisch-Ikegami, Bullington, Elgi, Epstien-Peterson, and Longley-rice[2-5]. Propagation models can be described or relegated to two distinct classes: deterministic and stochastic. The deterministic model is useful when mulitpath is caused by a small number of paths. The stochastic model is useful when multipath is caused by a large number of paths between the transmitter and receiver. In this study we found that the most suitable model for this study was the Hata-Okumura model. This model was designed for frequencies up to 2Ghz, and hence it is adapted in this research.

HATA-OKUMURA MODEL

The Hata-Okumura model is best suited for a large cell coverage (distances up to 100 km) and it can extrapolate predictions up to the 2GHz band [1]. This model has been proven to be accurate and is used by computer simulation tools. Here is the analytical approach to the model:

$$PL = 69.55 + 26.16 \log(f) - 13.82 \log(h_t) - a(h_m) + [44.9 - 6.55 \log(h_t)] \log(d) \text{ dB}$$

$$a(h_m) = [1.1 \log(f) - 0.7] h_m - [1.56 \log(f) - 0.8] \text{ dB} \qquad \text{for midsize city}$$

where f - operating frequency (MHz)

h_t - transmitting station antenna height (m)

h_m - mobile unit antenna height (m)

$a(h_m)$ - correction factor for mobile unit antenna height (dB)

d - distance from transmitting station

Using the following parameters: f = 2000 MHz, h_t = 40 m, h_m = 1.5 m, the loss predictions for this model is shown graphically in Figure 2.1.

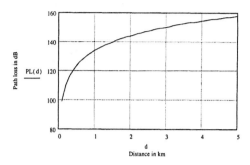

Figure 2.1 Hata-Okumura model

MATHCAD MODEL FOR POWER LOSS

To convert path loss data into a matrix format, the propagation coverage area is first viewed as a XY coordinate system. If we consider a transmitter as being located at the origin, then any position relative to this point whether occupied by a receiver or an obstruction can be defined in terms of x and y coordinates. For example, say a 5 × 5 km square map represents a propagation coverage area. For a 1 km grid resolution, x will take on a range of values such that $x = -5, -4.....5$ and similarly y such that $y = -5, -4.....5$. If obstructions are positioned at points (1, 4), (0, 4), (-2, 3) and (4, -2) respectively, then a mapped layout of the coverage area will be as shown in Figure 2.2.

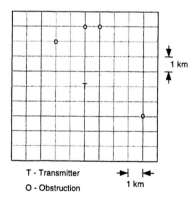

T - Transmitter
O - Obstruction

1 km

Figure 2.2 Mapped layout of propagation coverage area.

The function $fgrid(f,N)$ takes a function of two variables f as an argument and an integer N and returns a $(N+1)\times(N+1)$ grid of values of f over the square. Therefore, since a path loss equation is defined in terms of d, it is a function of x and y, and subsequently can be applied.

PLsim

$$
F(x,y) := \begin{vmatrix} f \leftarrow 1500 \\ h_t \leftarrow 40 \\ h_m \leftarrow 1.5 \\ m_c \leftarrow (1.1 \cdot \log(f) - 0.7) \cdot h_m - (1.56 \cdot \log(f) - 0.8) \\ d \leftarrow \sqrt{x^2 + y^2} \\ d \leftarrow 0.1 \text{ if } (x=0) \cdot (y=0) \\ PL \leftarrow 69.55 + 26.16 \cdot \log(f) - 13.82 \cdot \log(h_t) - m_c + (44.9 - 6.55 \cdot \log(h_t)) \cdot \log(d) \\ PL \end{vmatrix}
$$

Figure 2.3 Standard formula for Hata-Okumura model as a function of x and y.

After the program is executed, an 11 × 11 matrix M is returned as shown in Figure 2.4. However, since the Hata-Okumura model ignores the effects of buildings and streets, a few building obstructions are created. These simulated buildings each represent an assumed loss of 20 dB. A data matrix Z gives their exact locations with respect to the position of the transmitter (0, 0).

PLsim

M =

	0	1	2	3	4	5
0	159.688	158.205	156.807	155.618	154.802	154.509
1	158.205	156.354	154.509	152.842	151.628	151.175
2	156.807	154.509	152.055	149.624	147.663	146.876
3	155.618	152.842	149.624	145.996	142.485	140.818
4	154.802	151.628	147.663	142.485	135.639	130.46
5	154.509	151.175	146.876	140.818	130.46	96.054
6	154.802	151.628	147.663	142.485	135.639	130.46
7	155.618	152.842	149.624	145.996	142.485	140.818
8	156.807	154.509	152.055	149.624	147.663	146.876
9	158.205	156.354	154.509	152.842	151.628	151.175
10	159.688	158.205	156.807	155.618	154.802	154.509

Figure 2.4 Truncated 11× 11 matrix M.

The data matrix Z, as shown in Figure 2.5, actually represents simulated terrain data. However, in the real world of signal propagation this matrix will be much more detailed in terms of obstruction losses over an entire area. A typical urban environment will also require a grid resolution far greater than 1 km to simulate obstruction positions with some degree of accuracy.

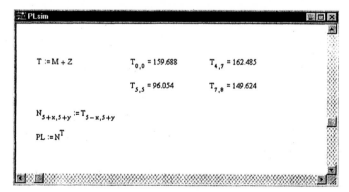

$Z =$

	0	1	2	3	4	5	6	7	8	9	10
0	0	0	0	0	0	0	0	0	0	0	0
1	0	0	0	0	0	0	0	0	0	0	0
2	0	20	20	0	0	0	20	20	20	0	0
3	0	0	0	20	0	0	20	20	20	0	0
4	0	0	0	0	20	0	20	20	20	0	0
5	20	0	0	0	0	0	0	0	0	0	0
6	0	20	20	0	0	0	0	0	0	0	0
7	0	20	20	0	0	0	0	0	0	0	0
8	0	0	0	0	0	20	0	20	20	0	0
9	0	0	20	0	0	0	0	20	20	20	0
10	0	0	0	0	0	0	0	20	20	20	0

Figure 2.5 Data matrix Z.

MATHCAD GENERATED SIGNAL LOSS CONTOURS

The addition of M and Z gives the total path loss matrix T. However, in order to display this result as a contour plot in its correct perspective, the T matrix needs to be fliiped around the x-axis, and its transpose is taken. These programming steps are shown in Figure 2.6.

$$T := M + Z \qquad T_{0,0} = 159.688 \qquad T_{4,7} = 162.485$$

$$T_{5,5} = 96.054 \qquad T_{7,8} = 149.624$$

$$N_{5+x,5+y} := T_{5-x,5+y}$$

$$PL := N^{T}$$

Figure 2.6 Programming steps to facilitate Mathcad default feature.

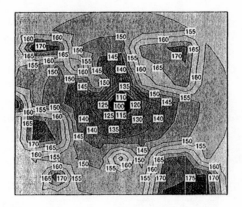

PL

Figure 2.7 Simulated PL contour plot for the Hata-Okumura model with added obstructions.

This figure shows the simulated PL contour plot for the Hata-Okumura model with added obstructions. Using the auto contour feature, Mathcad software displays actual path loss over the 10×10 km square simulated propagation area. The apparent elastic nature of the contour lines gives a good reproduction of signal propagation around obstructions. The numbered contour lines give an indication of the path loss value in a particular region relative to the center of the grid.

MATHCAD GENERATED BER CONTOURS

The signal-to-noise ratio can be expressed as a carrier power-to-noise ratio $P_rNR[6]$.

$$\frac{P_r}{N} = \frac{EIRP \, G_r/N}{L_s L} \tag{2.1}$$

where $EIRP$ is the transmitted signal power, and L_s is the path loss or space loss. For digital communication links it is common to replace noise power N with noise power spectral density N_o, given as

$$N_o = kT \tag{2.2}$$

where k is Boltzmann's constant (1.38×10^{-23} joule/K), and T the system effective temperature in degrees kelvin, which is a function of the noise radiated into the antenna and the thermal noise generated within the first stages of the receiver.

For BPSK modulation equation (2.1) becomes

$$\frac{P_r}{N_o} = \frac{S}{N_o} = \frac{E_b}{N_o} \cdot R_b \qquad (2.3)$$

where S is the average modulating signal power, E_b/N_o the bit energy per noise power spectral density, and R_b the bit rate.

The required $\dfrac{E_b}{N_0}$ for BPSK modulation is found to be given by:

$$\left(\frac{E_b}{N_o}\right)_{rec} = EIRP + G_r - R_b - kT - L_s - L \qquad (2.4)$$

The bit error probability for BPSK signaling is given by Sklar [7] as

$$P_b = Q\left(\sqrt{\frac{2E_b}{N_o}}\right) \qquad (2.5)$$

where the Q function is defined as

$$Q(x) = 0.5\left(1 - erf\left(\frac{x}{\sqrt{2}}\right)\right) \qquad (2.6)$$

The parameters used in plotting a BER curve are set as follows: P_t is the transmitting power given as 1 W ($EIRP = P_t \cdot G_t$), and G_t the transmitting antenna gain. Note here that $(E_b/N_o)_{rec}$, expressed as $EbNoR$ in the Mathcad program is defined as a function of PL over a range of 100 to 200 dB.

Bit error probability or bit error rate (BER) may be fully understood by considering the case of a digital communication system that has at its output, a sequence of symbols. The output of the system due to the influence of channel noise (which is assumed Gaussian) will be a different sequence of bits. In an ideal or noiseless system, both input and output sequences are the same, but in a practical system, they will occasionally differ. Therefore, the bit error probability may be defined as the probability that the input sequence of symbols is not equal to the output sequence of symbols. In a practical digital communication system, the values of bit error probability range from 10^{-4} to 10^{-7}. In practice, the bit error rate (BER) is used together with time intervals to provide performance objectives for digital systems, as stated in Townsend [6].

30

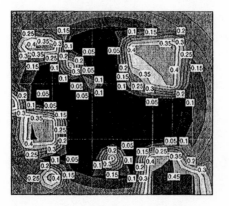

BER

Figure 2.8 Mathcad generated BER contour plot based on the Hata-Okumura model with obstructions.

EXPERIMENTAL PATH LOSS MEASUREMENTS

In developing a RF propagation model for a DS spread spectrum communication system, radio propagation measurements (at 2.4 GHz) were taken in Dade county, Miami. A particular area called the Ives Estates was targeted for adequate propagation measurements. This urban area was somewhat close to the Golden Glades interchange (the busiest intersection in the state of Florida). These linear propagation measurements of signal strength in decibel-milliwatts (dBm) vs. distance in meters (m) were integrated into a Mathcad program in order to generate a scatter plot used in its regression analysis.

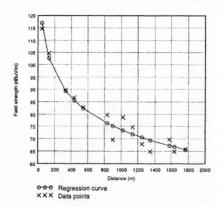

θ-θ-θ Regression curve
××× Data points

Fig. 2.9 Graphical result of the regression analysis.

For a greater degree of accuracy, the propagation coverage for the Ives Dairy Estates area is simulated using a 101 × 101 PL matrix. A matrix of this order results in a grid resolution of 1/10 km over a 5 × 5 km square map. Path loss values used in generating this matrix are obtained using the Mathcad program shown in Figure 2.10. This PL matrix as shown in Figure 2.10, gives standard path losses over the terrain as obtained by the Hata-Okumura based derivations.

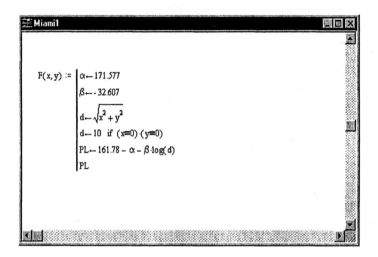

Figure 2.10 Mathcad program used in generating standard path loss values over the terrain.

A regression analysis was performed on the measured data resulting in a mean square error (MSE) of 10.75, an estimate of the error standard deviation σ, also called the standard error of estimate is given by

$$\sigma = \sqrt{MSE} \tag{2.7}$$

Based on a sample size of 10201, a vector of random numbers is generated from a normal distribution with zero mean μ and standard deviation σ, using a Mathcad built-in feature. The vector is then packed into an array of size 101 × 101. This new matrix B is then added to the original PL matrix M, which is defined as

$$M = fgrid(F,100) \tag{2.8}$$

Miami1

fgrid(F, 100) =

	46	47	48	49	50	51	52	53	54
44	83.394	82.37	81.536	80.984	80.79	80.984	81.536	82.37	83.394
45	81.711	80.385	79.259	78.486	78.208	78.486	79.259	80.385	81.711
46	79.956	78.208	76.628	75.478	75.048	75.478	76.628	78.208	79.956
47	78.208	75.882	73.578	71.721	70.974	71.721	73.578	75.882	78.208
48	76.628	73.578	70.141	66.813	65.233	66.813	70.141	73.578	76.628
49	75.478	71.721	66.813	60.325	55.417	60.325	66.813	71.721	75.478
50	75.048	70.974	65.233	55.417	22.81	55.417	65.233	70.974	75.048
51	75.478	71.721	66.813	60.325	55.417	60.325	66.813	71.721	75.478
52	76.628	73.578	70.141	66.813	65.233	66.813	70.141	73.578	76.628
53	78.208	75.882	73.578	71.721	70.974	71.721	73.578	75.882	78.208
54	79.956	78.208	76.628	75.478	75.048	75.478	76.628	78.208	79.956
55	81.711	80.385	79.259	78.486	78.208	78.486	79.259	80.385	81.711

Figure 2.11 PL matrix giving standard path loss values over the terrain.

The resulting matrix S, defined in Equation (2.9), is then modified to facilitate the set of random propagation measurements taken over the terrain as defined in Figure 2.12.

$$S = M + B \tag{2.9}$$

Miami1

$S_{52,54} := 88.974$ $S_{52,56} := 88.974$ $S_{52,58} := 83.974$

$S_{52,59} := 91.974$ $S_{53,60} := 86.974$ $S_{53,61} := 88.974$

$S_{53,63} := 96.974$ $S_{53,64} := 94.974$ $S_{52,64} := 96.974$

$S_{52,62} := 93.974$ $S_{52,60} := 94.974$ $S_{52,57} := 78.974$

$S_{53,59} := 79.974$ $S_{52,61} := 90.974$ $S_{52,63} := 97.974$

$S_{52,53} := 71.974$

Figure 2.12 Random propagation measurements defined as path loss values.

This matrix S, taken as the final path loss matrix is plotted as shown in Figure 2.13. This simulation of the propagation coverage gives an idea of the path loss in unmeasured areas as well, based on previous analysis. However, in the specific area in question, that is the Ives Dairy road, values of path loss derived based on the random propagation measurements were in the vicinity of those proposed by regression analysis. This indicates that the proposed propagation model is well within acceptable limits. However, recalling that all propagation measurements taken were based on line-of-sight, other propagation measurements are needed based on variable field parameters such as transmitting and receiving antenna heights and speed of the mobile unit.

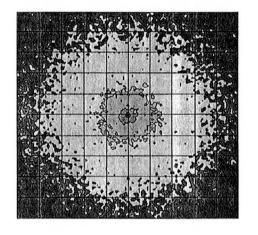

- >105dB

  90-105dB

- 75-90dB

- 60-75dB

- 45-60dB

PL

Figure 2.13 Propagation coverage simulation based on field measurements.

CONCLUSION

It is paramount when doing any communication system design that modeling of the RF propagation is accurate. For this cause, any propagation model should be optimized for its own particular environment. The prediction model presented based on the Miami propagation measurements, should be improved. Propagation measurements are needed for every facet of radio coverage, such as with variable transmitting and receiving antenna heights, with the effects of clutter, and the effects of buildings. As such the path loss constants obtained using regression analysis can be optimized, resulting in an established path loss empirical formula.

The Mathcad model has been shown to be practical method of simulating propagation coverage. However, if propagation coverage is limited to the microcell level, say 1 x 1 km square map, a 101 x 101 path loss matrix will result in a grid resolution of 1/50 km or 20 meters, producing far more accurate simulations. Overall, this Mathcad model approach can in the future provide a speedy and economic service to communications system design engineers.

LIST OF REFERENCES

1. T. S. Rappaport, "Wireless Communications: Principles and Practice," IEEE Press, Prentice-Hall, New Jersey, 1996.

2. B.H. Fleury and P.E. Leuthold, 'Radiowave Propagation in Mobile Communications: An Overview of European Research'. IEEE Comm. mag., Vol. 34, February 1996.

3. M. Hata, "Empirical Formula for Propagation Loss in Land Mobile Radio Services," IEEE Trans. on Veh. Technol., vol. VT-29, pp. 317-325, August 1980.

4. A. G. Longley and P. L. Rice, "Prediction of Tropospheric Radio Transmission Loss Over Irregular Terrain: A Computer Method," ESSA Tech. Rep. ERL 79 – ITS 67, US Govt. Printing Office, Washington DC, 1968.

5. J. Walfisch and H. L. Bertoni, "A Theoretical Model of UHF Propagation in Urban Environments," IEEE Trans. on Antennas Propagat., vol. 36, pp. 1788-1796, December 1988.

6. A. A. R. Townsend, "Digital Line-of-sight Radio Links: A Handbook," Prentice-Hall, London, 1988.

7. B. Sklar, "Digital Communications: Fundamentals and Applications," Prentice-Hall, New Jersey, 1988.

A Theoretical Analysis of Multiple Diffraction in Urban Environments for Wireless Local Loop Systems

Dave Crosby[1]*   Steve Greaves[2]   Andy Hopper[1]

[1] Laboratory for Communications
Engineering
Department of Engineering
University of Cambridge
Trumpington Street
Cambridge, UK

[2] Adaptive Broadband Limited
Westbrook Centre
Milton Rd
Cambridge, UK

Abstract

The simulation technique of Walfisch [1] is used to examine multiple diffraction in wireless local loop systems. The simulations results show that the average propagation characteristic is described by a two slope model. In the immediate vicinity of the basestation the propagation loss is found to have a distance dependence of 20 dB per decade. At greater distances the slope increases to approximately 40 dB per decade.

The distance at which the slope changes value is derived by considering the probability of Fresnel zone blockage.

1   Introduction

An wireless local loop (WLL) is a fixed radio communication system which delivers telephony and data services to the home or office in place of the traditional wireline network. In comparison to wireline networks, WLL's have lower capital costs and offer the potential for faster network deployment times. Figure 1 shows the typical arrangement of a WLL in which an elevated basestation is used to deliver telecommunication service over a radio channel to a subscriber's house. The subscriber's antenna is fixed and located in a high position that provides, where possible, a line-of-sight (LOS) communications path to the basestation. WLL's typically employ high gain, narrow beam antennas at both the subscriber's premise and basestation which serve to reduce transmit power requirements and minimise any interference between users.

---

*Contact: dbc20@eng.cam.ac.uk

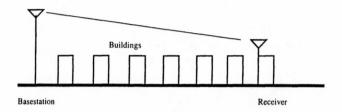

Figure 1: Wireless Local Loop

Like mobile systems, WLL's provide wide area coverage by reusing resources (e.g. frequency, time or codes) in cells which tessellate the geographic area. Coverage within each cell depends on the existence of a LOS path between the cell's basestation and the subscriber's antenna. Propagation within a cell is therefore characterised by free space transmission and is not affected by lognormal shadowing.

One consequence of reusing resources is that interference may occur between cells. The amount of downlink interference received by subscribers impacts directly on the capacity of a WLL system [2]. There have been two approaches to modeling downlink interference in the literature. Lee [2] takes a worst case analysis and assumes that any interference is LOS. While this may be reasonable for propagation over short distances, at longer distances the probability of a LOS path becomes less likely due to blockage by buildings and foliage. Consequently at longer distances interference will tend to be non-LOS (NLOS). This is the approach take by Gong [3], who assumed a 30 dB per decade law and included a log-normal shadowing factor.

In this paper we theoretically examine the propagation characteristics of the WLL channel by examining the influence of random building heights on the path loss for antennas elevated above rooftop. This analysis is based upon application of the Walfisch [1] model. We present a range of simulation results for various frequencies, basestation heights and building height distributions.

2 The Walfisch Model

Walfisch [1] proposed a semi-empirical model that is applicable to propagation in urban environments. This model assumes that a simple representation of a city, with the exception of the high rise core, is of equally spaced rows of buildings that are of uniform height. Propagation is then equated to the process of multiple diffraction past these rows of buildings.

To mathematically calculate the field at rooftop level each building is replaced by an absorbing half-screen and the Kirchoff-Fresnel equation numerically evaluated for an incident plane wave. The field at rooftop height is found to obey the following expression to within 0.8 dB:

$$Q(\alpha) = \begin{cases} 0.1 \left[\frac{\alpha}{0.03} \sqrt{\frac{d}{\lambda}} \right] & \alpha\sqrt{d/\lambda} \le 0.3875 \\ 1.0 & \text{otherwise} \end{cases} \tag{1}$$

where α is shown in Figure 2, d is the screen spacing and λ the wavelength. The field for receiver antenna heights below roof-line can be calculated by including a term to account for the diffraction over the final rooftop down to the antenna [1].

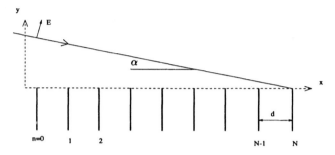

Figure 2: Plane-wave simulation geometry used by Walfisch

In obtaining (1), Walfisch simplifies the process of multiple diffraction by using a localised plane-wave approximation with angle of declination α for the spherical-wave radiation originating from the elevated basestation antenna. By comparing the results for a plane wave and cylindrical wave, Xia has since shown that at rooftop height, such an approximation is reasonable [4].

A number of authors [4] [5] have since developed theoretical solutions for the field at rooftop height. Nevertheless, the use of (1) is attractive from the point of view of its simplicity and its validity has been verified with measurements [6] [7] [8].

For application to WLL, it is necessary to evaluate the field above rooftop height. In this case, it has not been possible to find a theoretical expression for the field in the closed form of [4] and we have resorted to using the simulation technique of Walfisch.

3 Applying the Walfisch Model to WLL Systems

In this section we apply the Walfisch model to theoretically predict the effects of multiple diffraction on the WLL channel. We do not consider other factors such as foliage loss and terrain variation.

Our simulations are based on the cylindrical wave implementation of the Walfisch model [9]. Unlike the plane wave implementation, this requires explicit information regarding the location and height of the basestation, which acts as a line source. The resulting simulation geometry for the WLL scenario is shown in Figure 3. The x-axis is coincident with the average roof height. The basestation and subscriber antennas are positioned at (x, y) co-ordinates of $(0, h_t)$ and (R, h_s) respectively. Note that a negative value of h_s corresponds to the subscriber antenna being positioned below average roof level. We assign the coordinates (nd, h_n) to the top of the nth screen and constrain the subscriber antenna to a position vertically above each screen (i.e. $R = nd$).

Reflections from buildings behind the subscriber, which are included in the standard Walfisch model, are ignored in our case, as these will be rejected by a directional antenna.

It is usual to set the height of each knife-edge to the average building height, i.e. $h_n = 0\ \forall n$. However, at heights above average rooftop level we have found that the field is particularly sensitive to any variations in building height. Consequently in this analysis we assume building heights to be uniformly distributed between $-\delta_h/2$ to $\delta_h/2$ where δ_h is the difference between the maximum and minimum building heights. A similar approach was also used by Chrysanthos [10] for examining the influence of random building heights on the field amplitude at heights well below rooftop level.

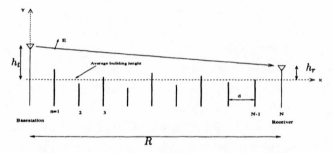

Figure 3: Simulation geometry for a cylindrically radiated wave

Following [9] we represent the radiation from the basestation as a cylindrical wave generated by a magnetic line source positioned at $(0, h_t)$ and oriented parallel to the z axis. The amplitude of the field in the plane of the first knife edge is therefore given by:

$$H_1(y) \approx \frac{e^{ikr}}{\sqrt{kr}} \qquad (2)$$

where $r = \sqrt{d^2 + (y - h_t)^2}$ and $k = 2\pi/\lambda$ is the wavenumber.

In general, the field above the $(n + 1)$th screen, $H_{n+1}(y)$, can be obtained from the field above the nth screen, $H_n(y)$, by applying the Kirchoff-Fresnel integral [11]:

$$H_{n+1}(y) \approx ke^{-j\pi/4} \int_{b_n}^{\infty} E_n(y')cos\theta_n \frac{e^{jkr}}{\sqrt{2\pi kr}}dy' \tag{3}$$

where $cos\theta_n = d/r$ and $r = \sqrt{d^2 + (y - y')^2}$. The above integral can be solved numerically by rewriting it as a summation of integrals over segments of size Δ and approximating the phase and amplitude of the integrand over each interval as linear functions[1].

To terminate the upper limit of the integration in (3) to a finite value we use a Kaiser-Bessel window function. In the notation of [9], we set the parameters of the Kaiser-Bessel function to ensure the field perturbation error [1] resulting from the windowing operation is less than ϵ. With reference to Equation (10) in [9], the resulting window parameters are:

$$y_t = h_s + (h_t + \delta_h/2))/2 + L/2\sqrt{\frac{(h_t - \delta_h/2)^2}{L^2} + \frac{\lambda}{2\pi\epsilon L}} \tag{4}$$

$$Y(x) = 0 \tag{5}$$

$$f_o = 15\sqrt{\lambda d} \tag{6}$$

In all our simulations we have used $\epsilon = 5 \times 10^{-3}$. Further details regarding the simulation procedure can be found in [9].

4 Simulation Results

A number of different systems were simulated. These are listed in Figure 4. Each simulation was performed over $N = 100$ screens, which for the given values of inter-screen spacing d, corresponds to a total transmission distance of between 2.5 to 5 km. Building height variations ranged from $\delta_h = 0$ to $\delta_h = 9$ metres. In order to obtain meaningful statistical measures we have repeated each simulation fifty times with a new sample of building heights used on each run. From the results the average and standard deviation of the field amplitude at heights above each screen could be calculated, allowing a path loss profile to be generated.

The simulation calculates the amplitude of the field at heights of $h_s = 0$ to $h_s = 5$ m in 0.5 m increments. The results in this section are presented in terms of diffraction loss. The diffraction loss in decibels for a subscriber antenna positioned above the nth screen at height h_s is defined as:

$$D = 20\,log_{10}(|H_n(h_s)|\sqrt{kr_o}) \tag{7}$$

Frequency (MHz)	d (m)	h_t (m)	δ_h (m)
600	50	10	3,7
600	25	20	3,7
2400	25	10	1,5,7
2400	50	10	1,3,5,7,9
2400	50	5	1,2,5,7,9

Figure 4: Parameters of the systems simulated

where $H_n(h_s)$ is the simulated field at height h_s in the plane of the nth screen and $r_o = \sqrt{(nd)^2 + (h_t - h_s)^2}$.

4.1 Average Diffraction Loss Characteristic

As an example, Figure 5 shows the diffraction loss for $h_t = 50$ m, $d = 25$ m, $\lambda = 0.125$ m and $\delta_h = 7$ m. In this Figure the small dot points are the diffraction losses calculated from fifty simulation runs and the circles represent the localised average of this data at distances of nd. For distances up to 400 metres, the average is observed to be approximately 0 dB as the probability of a LOS path is high and propagation is mostly free space (i.e. zero slope). At greater distances, the probability of a LOS path becomes less likely and the average of the diffraction loss exhibits a slope approaching 20 dB per decade. This type of behaviour was observed in all simulations where $h_s > 0$.

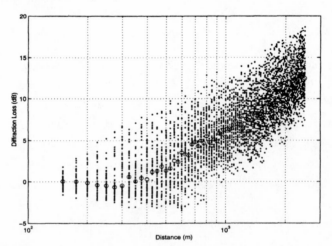

Figure 5: Simulated diffraction loss for $f = 600MHz$, $h_t = 20$ m, $h_s = 2$ m, $d = 25$ m, $\delta_h = 7$ m. The small dot points are the combined results from 50 simulation runs. The circles are the localised average loss at distances of nd, $n = 1, 2, \ldots$.

In Figure 6 we compare the average diffraction loss characteristic for different subscriber antenna heights. As the customer antenna is elevated, the chance of a LOS path increase and we find the distance at which the propagation characteristic changes from the free space to diffracted mode increases.

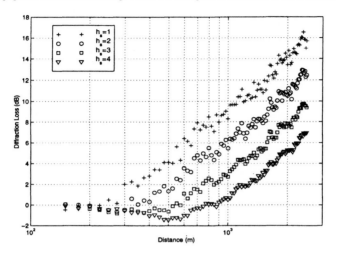

Figure 6: Simulated diffraction loss (localised average) for $f = 600MHz$, $h_t = 20$, $d = 25$ m, $\delta_h = 7$ and subscriber antenna heights of $h_s = 1, 2, 3$ and 4 m.

All our simulations revealed this type of behaviour. A simple expression for predicting the average diffraction loss at subscriber antenna heights $h_s > 0$ is therefore:

$$
L_d(R) = \begin{cases} 0 & R \leq R_o \\ 20 log_{10}(R/R_o) & \text{otherwise} \end{cases} \tag{8}
$$

where R_o is the break-point distance separating the LOS and diffracted regions. To apply the above equation it is necessary to relate R_o to parameters such as inter-screen spacing, basestation height, etc. One approach to predicting R_o is to consider the probability of buildings obscuring the mth Fresnel zone as the subscriber moves further away from the basestation. The probability of mth Fresnel zone blockage at a distance nd from the basestation is approximately:

$$
P_n = 1 - \prod_{i=1}^{i=n-1} \mathbf{P}\{h_i \leq f_m(id)\} \tag{9}
$$

where $\mathbf{P}\{E\}$ represents the probability of event E and $f_m(x)$ is the bound of the m_{th} Fresnel zone at distance x for a subscriber antenna positioned at (nd, h_s):

$$f_m(x) \approx h_t - x \tan\alpha - \sqrt{m\lambda \frac{x(nd - x)}{nd \cos\alpha}} \tag{10}$$

where $\alpha = atan((h_t - h_s)/nd)$.

The blocking probability P_n represents a cumulative distribution function (CDF) and has values $P_1 = 0$ and $P_\infty = 1$. The average distance at which blocking occurs is obtained by taking the expectation of the probability distribution function:

$$< R_o > = \sum_{n=2}^{n=\infty} \frac{(2n + 1)}{2} d[P_n - P_{n-1}] \tag{11}$$

In Figure 7 we compare the values of R_o obtained from simulation against those computed directly from (11) for a value of $m = 0.26$. Good agreement is observed although some over-prediction occurs at small distances.

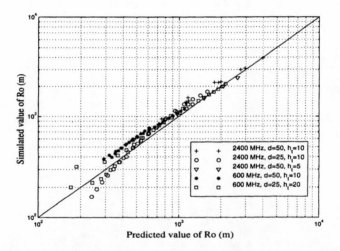

Figure 7: Comparison of the break-point distance as predicted by (11) and that obtained directly from simulation for the systems listed in Figure 4.

4.2 Probability Distribution Function

The above section has examined the average diffraction loss characteristic. In this section we examine the distribution of the diffraction loss data about this average.

Figure 8 shows a typical distribution function for a simulation with parameters $h_t = 10m$, $h_r = 3m$, $\lambda =$

$0.125m, \delta_h = 5m, d = 25m$. This is a plot of the percentage of diffraction loss data that have values less than abscissa, with the abscissa being measured relative to the localised average. The ordinate of this graph has been appropriately scaled such that a log-normal distribution appears as a straight line. Also shown for comparison are log-normal and uniform distributions with the same standard deviations and averages. The distribution function of the diffraction loss data is almost coincident with the curve for the log-normal distribution. In fact, the distribution function of the diffraction loss has maximum deviation of less than 1 % from the log-normal distribution. In comparison, the deviation from the uniform distribution is over 6 %. Consequently the diffraction loss is accurately modeled as having a log-normal distribution.

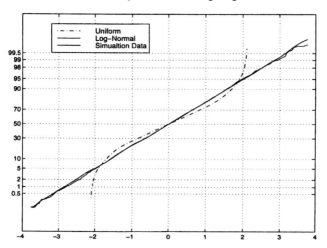

Figure 8: Probability distribution (solid curve) of the diffraction loss relative to the average diffraction loss for a building height variation of $\delta_h = 5m$. Also shown for comparison are the uniform and log-normal distributions (solid straight line) with the same average and standard deviation. Other simulation parameters were $h_t = 10m, h_r = 3m, \lambda = 0.125m, d = 25m$

5   Conclusions

In this paper we have applied the Walfisch simulation technique to examine multiple diffraction in WLL systems. In particular we have examined the case in which the subscriber antenna is at or above rooftop.

Our simulations have shown that the average path loss characteristic can be separated into a two regions. At distances close to the basestation the propagation is equivalent to free space. At larger distances the propagation loss approaches 40 dB/decade law. The break-point distance can be calculated by considering the probability with which the mth Fresnel zone is blocked. We have found that a value of $m = 0.26$

provides reasonably agreement with the simulation results.

The distribution of the diffraction loss about the average was shown to be approximately log-normal. This is in agreement with the generally accepted view in the literature.

This paper has not considered other factors, such as the presence of foliage and differences in building design and construction. In some environments these factors may significantly affect the propagation characteristic.

References

[1] J. Walfisch and H. Bertoni. A theoretical model of UHF propagation in urban environments. *IEEE Trans. Antennas Propagat.*, 36(12):1788–1796, 1988.

[2] W. Lee. Spectrum and technology of a wireless local loop system. *IEEE Personal Communications*, Feb. 1988.

[3] S. Gong and D. Falconer. Cochannel interference in cellular fixed broadband access systems with directional antennas. *Wireless Personal Communications*, 1999.

[4] H. Xia, H. Bertoni, L. Maciel, A. Lindsay-Stewart, and R. Rowe. Radio propagation characteristics for line-of-sight microcellular and personal communications. *IEEE Tran. Antennas and Propagat.*, 41(10):1439–1447, 1993.

[5] S. Saunders and F. Bonar. Prediction of mobile radio wave propagation over buildings of irregular heights and spacings. *IEEE Trans. Antennas and Propagat.*, 42(2):137–144, 1994.

[6] P. Eggers and P. Barry. Comparison of a diffraction based radio propagation model with measurements. *Electronic Letters*, 26(8):530–531, 1990.

[7] K. Low. A comparison of CW-measurements performed in Darmstadt with the COST-231-Walfisch-Ikegami model. Technical report, Damstadt, Germany, Sept 1991.

[8] Cost: Urban transmission loss models for radio in the 900 MHz and 1800 MHz bands. Technical report, The Hague, The Netherlands, Sept. 1991.

[9] H. Bertoni L. Piazzi. Effect of terrain on path loss in urban environments for wireless applications. *IEEE Trans. on Antennas and Propagat.*, 46(8):1138–1146, Aug. 1998.

[10] C. Chrysanthou. Variability of sector averaged signals for UHF propagation in cities. *IEEE Trans. Vehic. Tech.*, 39(4):352 – 358, Nov. 1990.

[11] M. Born and E. Wolf. *Principle of Optics*. Pergamon Press Ltd, Oxford, 5 edition, 1975.

Active microstrip antenna for personal communication system

Marian Wnuk, Władysław Kołosowski, Marek Amanowicz, Tomasz Semeniuk
Military University of Technology, Electronics Faculty
str. Kaliskiego 2, 01-489 Warsaw, Poland
phone: (+48 22) 685-92-28   fax: (+48 22) 685-90-38
E-mail: mwnuk@wel.wat.waw.pl.

Abstract - Intensive development of cellular personal communications system has been observed lately. Thus, protection of a man, and especially protection of his head against non-ionizing electromagnetic radiation generated by cellular telephones is becoming one of the most important problems. The results of elaborated microstrip antennas which have minimized radiation towards the user's head are presented in this paper.

1. Introduction

In portable cellular personal communication devices which are used at present, a considerable part of radiation energy (up to 45 %) is absorbed by the user's head. It may have a harmful effect on his health.

Fig.1. Omnidirectional handset antenna radiation patterns

Therefore, protection of a man against radiation of radio communications system is an important problem. Protection from this radiation may by carried out on the basis of two principles:

- limitation of electromagnetic fields power emitted towards the user's head to the necessary minimum.
- limitation of the time for people staying in these electromagnetic fields.

48

The second principle that is, the length of a telephone call time, depends mainly on the speaker itself.

The first principle concerning the limitation of electromagnetic power absorbed the user's head may be based on changing of the radiation pattern which can be obtained by using a new type of antenna. The quarter-wave dipole which has been used so far, has an omnidirectional pattern

(plane $\vec{H}$).

2. Requirements for the antenna radiation pattern

Actually, there is lack of formal requirements (or practically implemented solutions) for the desired radiation pattern. The matter is complicated by the fact that the user's head is in the area of the near zone antenna. Therefore, it is necessary to find a compromise between the requirements for the availability of signals received by the antenna from all directions on the one hand , and the protection of a human head from radiation on the other hand.

We assume that the radiation pattern in the plane $\vec{H}$ should be defined as it is shown in fig.2, It is assumed that in the vector of $\vec{H}$ plane the radiation level in the whole area, except the area defined by the head protection angle within the range of 360^{0}, should be uniform. The problem of reverse radiation in disputable.

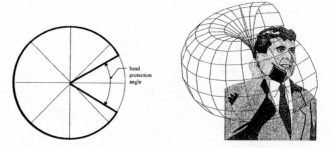

Fig..2. Requirements for antenna radiation pattern

On the one hand it is necessary to receive the signals emitted by a base station located in the operator head direction, but then the human head, especially some of its elements like bones, brain and skin which are characterized by high level of thermal conductivity (14.6, 8.05,

4.42 mW/cm^2°C respectively), should be exposed to transmitter radiation with minimal radiation power.

The radiation pattern shown in fig.2 has been accepted for practical analyses of the designed antenna systems.

3. Modelling of microstrip antennas

The thorough analysis of microstrip antennas which takes into account the structure of the layer and which is true for each frequency, is based on Green function and moment method. This method is based on solving the integral equation concerning the electric field generated by the currents flowing in the antennas element and its feeding systems. These currents are unknown. We simulate the flow of inducted current by means of distribution for base and test currents, next we test their mutual reaction by means of the functions. According to [L- 7] the reaction has the form of:

$$< \vec{E}^i(\vec{J}_m); \vec{J}_i > = -\sum_n I_n < \vec{E}^s(\vec{J}_m); J_n > \tag{1}$$

The unlimited sequence of these functions is necessary for exact solution. We assume the limited number of these functions and, thus we obtain approximate solution. The mutual reaction of the whole analysed system can be expressed in the form of a matrix equation:

$$[z_{mn}][I_n] = [V_m] \tag{2}$$

By solving this equation we define the distribution of the currents flowing along the analysed structure on condition that the elements of general matrix impedance, which in our case has the form of:

$$z_{mn} = -\frac{1}{\pi^2} \int_0^\infty \int_0^\infty F\left[G(k_x,k_y)\right]\left\{F\left[J_0(k_x)\right]\right\}^2 \cdot$$
$$\cos[k_x(x_m - x_n)] \cdot \cos[k_y(y_m - y_n)dk_x dk_y \tag{3}$$

where: $F\left[G(k_x,k_y)\right]$ - Fourier transform of Green function

$F\left[J_0(k_x)\right]$ - Fourier transform of base current

x_n, y_n, x_m, y_m - respectively, the coordinates of the situated means of base and testing

functions.

With the defined current distribution we can express the radiation pattern in the dipole plane by the following equation:

$$\vec{E}(\varphi) = \sum_{n=1}^{N} I_n \left[\vec{E}(\vec{J}_n) \right] \qquad (4)$$

where: I_n - coefficient of current distribution

The microstrip antennas patterns (presented in fig. 9a, 10a,) have been calculated on the basis of these relationships

4. Microstrip dipole antenna

Microstrip dipole antenna has been designed for GSM hand-held unit which operates at 900 MHz band. It can also be easily implemented for DCS system which operates at 1800 MHz. The structure of the dipole antenna with coaxial feeding is presented in figure 3. The microstrip dipole radiator is excited by the microstrip resonator which is coupled with the handset input.

There are two possibilities of feeding the dipole antenna i.e. using coaxial feedline (fig.3) or unsymmetrical stripline - USL (fig.4)

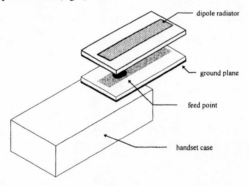

Fig.3. Microstrip dipole antenna with coaxial feedline

Typically, the microstrip radiator is excited directly from the feedline when USL is used for feeding the antenna, as it is shown in figure 4.

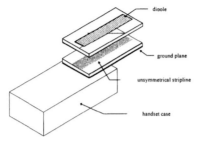

Fig.4. Microstrip antenna with unsymmetrical stripline

The multilayer technology used for dipole antennas is described here. This structure was selected to obtain the necessary bandwidth of the antennas (about 10%) which is necessary for GSM application.

Taking into account the requirement for the small size antenna for GSM application two types of multilayer microstrip dipole antennas were manufactured and tested i.e. :

- open-circuit half-wave dipole,
- short-circuit quarter-wave dipole.

The structure of these antennas is presented in figure 5.

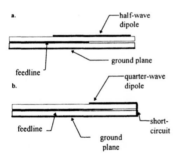

Fig.5. Half-wave (a) and quarter-wave (b) dipole antennas

Special attention should be paid to a quarter-wave dipole antenna due to its size which is especially important for 900 MHz band application. This antenna is approximately half the size of the half-wave antenna. It can be expected that the radiation pattern of a quarter-wave antenna in the E-plane may be sufficiently wide to achieve the optimal values of the antenna parameters.

52

5. Microstrip patch antenna

The application of a patch antenna for mobile communications is possible when higher frequency bands are considered (e.g. f > 1 GHz). The structure of a multilayer microstrip patch antenna is presented in figure 6. The upper rectangular patch radiator of L in length and W in width is excited by a slot placed on the upper side of the lower layer of the antenna. This slot is coupled with unsymmetrical feedline. The dimensions L and W as well as the slot location were selected empirically in accordance with the bandwidth criterion.

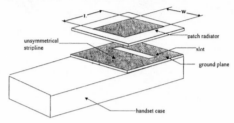

Fig.6. Structure of patch microstrip antenna

Next we optimize the dimensions of the patch that is its resonance length and the length of the slot so that the real part of the input impedance and the wave impedance of the feeding line should be equal. In case of a multilayer structure the process of designing is more complicated due to greater manipulation freedom. The requirement for a small size of a handset antenna makes this structure effective at a higher frequency band (e.g. for DCS 1800 system).

6. Measurements results

The construction of a microstrip antenna on a multilayer dielectric is presented in fig.7.

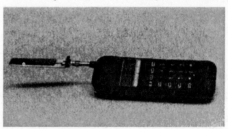

Fig.7. Microstrip dipole antenna with GSM handset

The empirical verification of dipole and patch microstrip antennas characteristics was performed. Measurements were made in free space and in user presence to investigate the influence of handset antenna radiation on the user's head. During the experiments the user was standing on rotary platform and was holding GSM handset at 45° to the horizon. Some results of quarter-wave dipole antenna measurements are presented in figure 9 i.e.: standing wave ratio, radiation pattern in free space. The similarly measured characteristics of microstrip patch antenna are presented in figure 8.

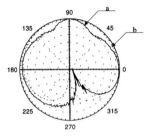
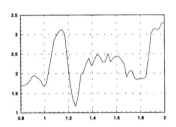

a. Radiation pattern in free space

b. Theoretical characteristic

c. - Standing wave ratio

Fig.8. Patch antenna measurements results

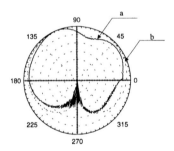
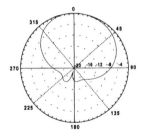

a. Radiation pattern in free space

b. Theoretical characteristic

c. Radiation pattern with man presence

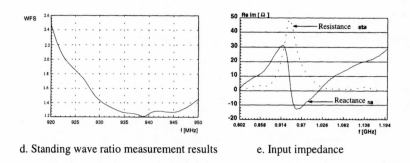

d. Standing wave ratio measurement results   e. Input impedance

Fig.9. Dipole antenna measurements results

7. Microstrip active antenna

Another method of reducing the dose of radiation absorbed by the user is to lower the intensity of the electromagnetic field around the exposed living tissue. To achieve this goal without decreasing the range, the effectiveness of the receiving antenna has to be increased. In case of mobile terminals the use of larger antennas is ruled out but a solution may be to use active antennas.

The useful signal power at the antenna output is a function of many variables

$$P_R = f[P_T, G, S_{eff}(K), L_s, L_p]$$ (5)

where: P_T - radiated power

        G - transmitter antenna gain

        S_{eff} - effective aperture

        K - amplifier gain

        L_s - system losses

        L_p - propagation losses

When an 18dB amplifier is used, the maximum distance between the antennas can be increased eight times. If the distance remains unchanged, power can be reduced sixty four times. In effect the life of the mobile phone battery is prolonged and the dose of energy absorbed by the user is dramatically reduced.

The next step was to analyse an array consisting of a microstrip antenna in series connection with the UTO 1002 and UTO 1054 amplifiers. Results of the measurements conducted are shown in Fig.10.

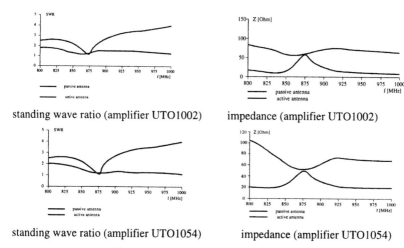

standing wave ratio (amplifier UTO1002)   impedance (amplifier UTO1002)

standing wave ratio (amplifier UTO1054)   impedance (amplifier UTO1054)

Fig.10. Measurement results of dipole antenna

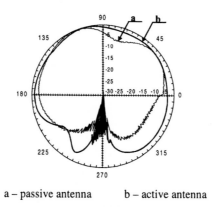

a – passive antenna   b – active antenna

Fig.11. Measured radiation pattern in free space

An analysis of the measurement results indicates that radiation towards the user's head was dramatically reduced, as shown in Table I.

Table I.

Class of mobile station	Passive antenna	Active antenna	Limited power density
GSM 900 (0.8 W) Class 5	0.004 W/m^2	0.00006 W/m^2	0.1 W/m^2
GSM 900 (2 W) Class 4	0.001 W/m^2	0.00017 W/m^2	

8. Conclusion

The results of the research show that the antennas presented here can be used in mobile telephones working in the 900 MHz band. Dipole antenna is preferable to GSM 900 applications while patch antenna may be used effectively when cellular system operates at the frequencies above 1 GHz. Both model antennas are characterized by reduced radiation towards the user's head. The next step is to make a casing for the antenna, which would not distort the radiation pattern.

References

[1] M. Amanowicz, W. Ko³osowski, M. Wnuk, A. Jeziorski „Microstrip antennas for mobile communications'' Proc. Of the Conference VTC'97 Phoenix , May.1997, USA.

[2] Bahl I.J Microstrip antennas with paper-think dimensions. Microwaves No.10.1979

[3] Guy A.W, Lehmann J.F, Stonebridge J.B. Therapeutic applications of electro-magnetic power. Proceedings of the IEEE vol. 62 No. 1 1974

[4] Jensen M.A, Rahmat-Samii Y, Performance analysis of antennas for hand-held transceiver using FDTD IEEE Transactions on antennas and propagation. Vol 42 No 8.

[5] R.J.Mailloux, J.F.McIlvenna, N.P.Kernweis: Microstrip Array Technology, " IEEE Trans Antennas and Propagat. AP-29, No.1, January 1981.pp.25-38

[6] Xiao-Hai Shen,A.E. Vandenbosch, A.R.Van de Capelle: Study of Gain Enhancement Method for Microstrip Antennas Using Moment Method, " IEEE Trans. Antennas and Propagat. AP-43 No 3, 1995.

[7] W. Kołosowski, M. Wnuk „Impedancja wzajemna anten liniowych umieszczonych na podłożu dielektrycznym'' Biuletyn WAT 1988 Nr 11.

Co-located, Dual-band, Multi-function Antenna System for the GloMo Universal Modular Packaging System

J. S. McLean, J. LaCoss[1], J. R. Casey[2], E. Guzman, G. E. Crook, and H. D. Foltz

The University of Texas–Pan American
Department of Electrical Engineering
1201 West University Drive
Edinburg, TX 78539
e-mail: mclean@ccsi.com

Abstract

The GloMo Universal Modular Packaging System is an ultra-high-density handheld system for mobile computing and communications. Typically, such a system supports several different radios. In particular, one version supports a 2450 MHz ISM band spread-spectrum WLAN as well as an AMPS phone/modem. Another version supports 915 MHz and 2450 MHz WLANs. Aside from supporting multiple radios, the antenna for this system must maintain the high packaging density the system; that is, it must occupy a small volume. Furthermore, it must be mechanically and environmentally robust and therefore suitable for the military and law enforcement applications for which the system is intended. In keeping with the high-density packaging philosophy, the antenna must serve as an effective heat sink for the internal microprocessor as well as the digital radio. Finally, because of the multipath interference generally present in the UHF radio spectrum at low antenna heights and because of the random nature of the orientation/positioning of any handheld radio, the antenna system is required to support diversity reception.

The antenna system presented here is a co-located combination of two types of radiating elements. One element is a heavily top-loaded, asymmetric, shunt-tuned monopole; alternatively, it could be classified as variation of a planar inverted-F antenna. The shunt tuning post is large in cross section and serves as a thermal path to allow the top-loading plate of the monopole to serve as an effective thermal radiator for the microprocessor and other heat-generating devices in the radio system. The other radiating element is a dual-polarization patch antenna which is fed coaxially through the shunt grounding post of the monopole. This feed arrangement provides isolation between the elements. The feed arrangement incorporates transmission line transformers which match the patch antenna over the entire 2450 MHz ISM band. In one configuration, the asymmetric monopole is designed to cover the 824-890 MHz AMPS cellular telephone band. In another, it is tuned to cover the 902-926 MHz ISM band.

1 GloMo Universal Modular Packaging System

The GloMo Universal Modular Packaging System (GUMPS) is a concept for a robust handheld data terminal with ultra-high packaging density, developed under the Global Mobile Information Systems

---

[1]USC/Information Sciences Institute, 4676 Admiralty Way, Marina del Rey, CA 90292

[2]University of Wisconsin–Madison, Department of Electrical and Computer Engineering, 1214 Engineering Drive, Madison, WI 53706

Project. The aim of the effort is to develop a robust, untethered data communication node capable of exploiting multiple, diverse communications channels in order to provide connectivity under adverse conditions. The intended applications originally included military and law enforcement scenarios, but the design concepts have been shown to be applicable to numerous commercial applications. The high packing density, along with the large number of supported RF devices combine to make design of the antenna system challenging. Among the RF devices supported by the GUMPS system are:

- 915-926 MHz ISM band WLAN

- 2400-2500 MHz ISM band WLAN

- AMPS telephone/MODEM

The AMPS telephone/MODEM will likely be supplanted by another (possibly 2) radio. However, this radio will most likely also operate in the 800-1000 MHz frequency range; possibilities include a GSM mobile telephone and a 915 MHz ISM band WLAN. In any case, what is required is a multi-function aperture which occupies minimal volume and is mechanically robust. At the time of the writing of this paper, the operating band with the lowest center frequency is the AMPS cellular telephone band which extends from 824 to 895 MHz. The wavelength at the lower end of this frequency range is about 33 cm. Thus, the antenna which covers this band is required to be *electrically-small*, that is, having its largest dimension small compared to a wavelength. Furthermore, it is required to be extremely low-profile, that is, one dimension is required to be much less than a wavelength. Bandwidth and electrical efficiency limitations imposed by fundamental physics [1, 2] greatly complicate the design of such antennas.

One thrust of this effort is to examine the trade off between a multi-band, single-port antenna and a set of co-located antennas. The former may offer some size and complexity advantages but requires an external multiplexer. The co-located system requires no external multiplexer as it provides distinct ports for each band. However, it is in most cases not possible to provide the band-to-band isolation available with a multiplexer. Thus some external filtering is required if moderate-power transmitters are involved.

In addition, the antenna system for the GloMo Universal Modular Packaging System is one for which packaging considerations are of paramount importance. While electrical performance must be maintained, packaging constraints cannot be violated. Our approach to this problem consists of a dual-polarization microstrip patch antenna for the 2.45 GHz band co-located with a capacitively loaded planar inverted-F antenna (PIFA) used for one of the lower frequency bands.

2 Capacitively-Loaded, Diagonally-Fed Planar Inverted-F Antenna

2.1 Inverted-F Geometry

The planar inverted-F antenna [3] (PIFA) can be viewed in two distinct ways: as an extreme case of an asymmetric top-loaded, shunt-fed monopole, or as a derivative of the quarter-wave microstrip patch. In a symmetric wire monopole, vertically-directed current provides an azimuthally symmetric pattern. Addition of symmetric top-loading, for example a capacitive disk or set of radials, creates horizontal currents whose contributions tend to cancel in far field, so that the pattern remains vertically polarized and azimuthally symmetric. In an asymmetrically loaded antenna such as the inverted-L [4, 5], the radiation from the horizontal current in the top section does not cancel, thus possibly providing enhanced radiation in addition to height reduction (for resonance) at the expense of a distorted radiation pattern and cross polarization. The top conductor in the inverted-L antenna can take on a planar or tapered geometry providing enhanced bandwidth over a wire inverted L. When the length of the antenna is significant fraction of a wavelength but the height is not, the antenna takes on the characteristics of a lossy transmission line resonator as shown in Figure 1. This transmission line resonator may be tapped at some mid point in order to scale the impedance level to a higher value allowing a direct match to a 50 Ohm system. Finally, capacitive tuning may be added at the open circuit end and at the input [6] as shown in Figure 2. In a PIFA configuration the antenna becomes three dimensional, with the top conductor planar.

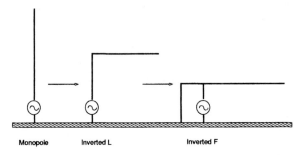

Figure 1: **Derivation of Inverted-F Geometry**

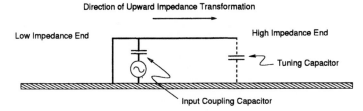

Figure 2: **Schematic Representation of Hybrid, Shunt-tuned Antenna Element**

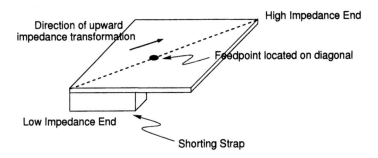

Figure 3: **Hybrid, Shunt-tuned Antenna Element**

2.2 Hybrid Shunt-tuned Antennas

The present design is a derivative of the inverted-F antenna in that it is essentially a tapped resonator. Unlike the wire inverted-F antenna, the resonator is three-dimensional. It is somewhat

Figure 4: **Package Design Showing Grounding Strap in Foreground**

Figure 5: **Package Design Showing Tuning Capacitor in Foreground**

like a quarter-wave resonator being shorted at one end and open at the other. However, the resonator can also be thought of as a shunt-fed, top-loaded asymmetric monopole. The arrangement is shown in Figure 3. This design retains the advantages of the element shown in Figure 2: the resonance frequency of the resonator can be adjusted via the capacitance at the high impedance end and the overall impedance level can be adjusted by repositioning the tap or feed point as shown. This flexibility is essential in that the packaging constraints allow little flexibility in the size or shape of the antenna. Photographs of the antenna package mounted on a mockup of the GUMPS unit are shown in Figures 4, and 5. The impedance locus for the top-loaded, asymmetric, shunt-tuned monopole is shown in Figure 6. As can be seen the locus encircles the origin thus providing broadband operation. The second, smaller knot in the locus is due to unavoidable package effects. The input return loss and radiation patterns are shown in Figures 7, 8, and 9.

3   Dual-polarization 2450 MHz ISM band Microstrip Patch Design

The dual-polarization microstrip patch is implemented on a soft, PTFE-based substrate ($\epsilon_R = 2.2$). In order to accomodate the package geometry in which the feedthroughs are routed through the grounding strap on the low frequency antenna, the patch is edge-fed. Thus, a relatively high feedpoint impedance is encountered, approximately 200 Ohms. Therefore, quarter-wave microstrip transmission line transformers with characteristic impedances of 100 Ohms and quarter wave frequencies of 2450 MHz were implemented between the coaxial feedthroughs (these had characteristic impedances of 50 Ohms) and the feedpoints on the patch antennas. The geometry of the dual polarization patch antenna including transmission line transformers and feedthroughs is shown in Figure 10 and 11. Unfortunately, the quarter-wave transformers have the disadvantage of an intrinsic narrowbanding effect; that is, they tend to store energy of the same form as the antenna at any given frequency. To understand this, consider the input impedance of a transmission line of characteristic impedance Z_0, phase velocity c, and length l terminated in complex impedance Z_l:

$$Z_{in} = Z_0 \frac{Z_l \cos(\beta l) + j Z_0 \sin(\beta l)}{Z_0 \cos(\beta l) + j Z_l \sin(\beta l)} \tag{1}$$

The frequency slope of the input impedance is given by:

$$\frac{\partial Z_{in}}{\partial \omega} = Z_0 \frac{\left[\frac{l}{c}\right]\left[-Z_l \sin(\beta l) + j Z_0 \cos(\beta l)\right] + \frac{\partial Z_l}{\partial \omega} \cos(\beta l)}{Z_0 \cos(\beta l) + j Z_l \sin(\beta l)}$$
$$- Z_0 \frac{Z_l \cos(\beta l) + j Z_0 \sin(\beta l)}{(Z_0 \cos(\beta l) + j Z_l \sin(\beta l))^2} \left\{\left[\frac{l}{c}\right]\left[-Z_0 \sin(\beta l) + j Z_l \cos(\beta l)\right] + j \frac{\partial Z_l}{\partial \omega} \sin(\beta l)\right\} \tag{2}$$

At $\beta l = 90°$,

$$\left.\frac{\partial Z_{in}}{\partial \omega}\right|_{\beta l = 90°} = j Z_0 \left[\frac{l}{c}\right]\left[1 - \frac{Z_0^2}{R_l^2}\right] - \frac{Z_0^2}{R_l^2} \left.\frac{\partial Z_L}{\partial \omega}\right|_{\beta l = 90°} \tag{3}$$

since

$$Z_l|_{\beta l = 90°} = R_l \tag{4}$$

The microstrip patch antenna is operated at its lowest parallel resonance. Hence, it is reasonable to model the the load as a lumped parallel resonant circuit over the frequency range near this resonance. Thus we expect a negative reactance slope at resonance.

$$\left.\frac{\partial Z_{in}}{\partial \omega}\right|_{\beta l = 90°} = j Z_0 \left[\frac{l}{c}\right]\left[1 - \frac{Z_0^2}{R_l^2}\right] + j \frac{Z_0^2}{R_l^2} \frac{2 C_l}{R_l^2} \tag{5}$$

Since

$$Z_0 = \sqrt{Z_{0system} R_l} \tag{6}$$

Because $\sqrt{Z_{0system}}$ is 50 Ohms, Z_0 is necessarily less than R_l. Therefore the reactance slopes add. Compensation could have been provided using a stub matching network. However, this approach was not used for two reasons:

1. Geometrical constraints greatly complicate the layout of such a network, and

2. The non-progressive nature of energy flow in such a matching network would exacerbate coupling to the co-located monopole element. That is, a stub matching network would couple more strongly to external fields than would a quarter-wave transformer. Because of the extreme proximity of elements in this antenna system, coupling between elements is critically important.

Nevertheless, the impedance bandwidth provided by the edge-fed patch with quarter-wave transformers is more than adequate for the ISM band radio. The input return loss at the two input ports is shown in Figure 12, and the E-plane and H-plane patterns are shown in Figure 13.

4   Co-location

Plots of the isolation between the two ports of the microstrip patch antenna, and of the isolation between the PIFA antenna and the microstrip patch, are shown in Figure 14. While the isolation obtained is acceptable for the intended operation, it is interesting to note that the coupling between the AMPS band PIFA and the ISM band patch is stronger at the 850 MHz PIFA resonance than at the 2.45 GHz patch resonance.

5   Diversity Operation

The diversity action of the dual polarization patch antenna can be quantified with the computation of the inner product of the two far field patterns. However, an approximation to the correlation coefficient can be obtained from the normalized mutual resistance between the two ports [7], under the assumption that the most of the antenna power pattern is in directions which correspond to likely angles-of-arrival for incoming waves. The correlation computed from the isolation and input impedance data presented earlier is 0.16 to 0.18 in the center region of the ISM passband.

6   Thermal Design and Analysis

Heat sinks provide thermal dissipation through convection, conduction, and radiation. The dielectric substrate of the microstrip patch partially insulates the top side of the PIFA top plate, so that heat loss is primarily through the ground plane and the underside of the top plate. The grounding strap on the PIFA is crucial transferring heat from the base to the top plate. The thermal resistance of the strap defined by

$$\theta = \frac{\Delta T}{P}$$

where ΔT is the temperature differential in degrees Celsius and P is the power flow in Watts is given by

$$\theta = \frac{l}{\kappa A} \; (\tfrac{^\circ \mathrm{K}}{\mathrm{W}}) \tag{7}$$

where κ is the thermal conductivity, l is the length of the path, and A is the cross sectional area of the path. As the grounding strap is composed of aluminum, the thermal conductivity is 2.3 $\frac{\mathrm{W}}{\mathrm{cm}-^\circ \mathrm{K}}$ and the thermal resistance of the strap is .059 $\frac{^\circ \mathrm{K}}{\mathrm{W}}$.

Of course, a quantitative thermal analysis would require a numerical simulation including the effects of radiative and convective heat transfer to the ambient. However, the main goal here was to develop an antenna geometry which would allow effective heat transfer from the ground plane to the top plate. The dual polarization patch antenna severly inhibits heat transfer from the top plate to the ambient. However, other designs involve the use of the PIFA element alone. In this case, the top plate could effectively serve as a thermal radiator. One such design involves a second PIFA element (without the patch elements) to provide diversity action in the 800-900 MHz range.

Because both the patch and the PIFA antennas have bandwidths just sufficient for the application requirements, it is critical that the frequency responses not shift excessively with rising temperature. The PIFA resonant frequency is controlled primarily by the lateral dimensions of the aluminum top plate, and secondarily by the RT-duroid load capacitor. The microstrip patch resonance is also determined by its lateral dimensions, which will expand with the underlying RT-duroid substrate. If we take the 1 dB additional mismatch loss at the band edges as the acceptable limit, the PIFA can tolerate approximately 15 MHz shift in resonant frequency, while the patch can tolerate approximately 20 MHz. Taking the thermal expansion of aluminum at 25 $\frac{\mathrm{ppm}}{^\circ \mathrm{C}}$ and that of the RT-duroid at 90 $\frac{ppm}{°C}$, a 25 °C heat rise will shift the resonance frequency of the PIFA about 500 kHz, and the patch about 5 MHz.

The capacitive coupling at the input provides extremely effective rejection in the PIFA element at frequencies well below 800 MHz. This is very important when the PIFA is used as a heatsink because is greatly reduces the conducted emissions path which would exist between the heatsinked components and the 800-900 MHz input port.

7 Conclusions

The culmination of this effort was a compact, robust, multi-function antenna/aperture which allows simultaneous operation of two or more radio systems with minimal co-site interference. One prototype was successfully demonstrated with a high-power (3 Watt) AMPS telephone and a 2.45 GHz WLAN operating simultaneously. The geometry of the antenna system naturally lends itself to use as an effective heat sink and thermal radiator thus providing an additional measure of packaging flexibility. Effective diversity operation is implemented for the 2.45 GHz system providing some mitigation of multipath fading and allowing re-orientation of the unit by the user. Finally, the geometry of this antenna system may serve as a base for more complex co-located antenna systems.

8 Acknowledgements

This work was supported by the DARPA Global Mobile Information Systems Program under Contract DABT63-97-C0041. DARPA assumes no responsibility for the contents of this paper.

References

[1] R. C. Hansen, "Fundamental Limitations in Antennas", *Proc. of IEEE*, Vol. 69, No. 2, Feb. 1981.

[2] J. S. McLean, "A Re-examination of the Fundamental Limits of the Radiation Q of Electrically-small Antennas", *IEEE Trans. Ant. Prop.*, vol.44, no.5, May 1996, pp. 672-676.

[3] T. Taga and K. Tsunekawa, "Performance Analysis of a built-in planar inverted-F antenna for 800 MHz band portable radio units," IEEE J. Select. Areas Commun., vol.SAC-35, pp.921-929, June 1987.

[4] W. L. Weeks, *Antenna Engineering*, McGraw-Hill Book Company, New York, 1968, pp 27-56,

[5] R. W. P. King and C. Harrison, "Transmission Line Missile Antennas," *IRE Trans. on Antennas and Propagation*, vol AP-8, Jan 1960.

[6] C. R. Rowell and R. D. Murch, "A Capacitively Loaded PIFA for Compact Mobile Telephone Handsets," *IEEE Trans. Ant. Prop.*, vol.45, no.5, May 1997, pp.837-842.

[7] R. G. Vaughan and J. B. Anderson, "Antenna Diversity in Mobile Communications," *IEEE Transactions on Vehicular Technology*, Vol. VT-36, No. 4, November 1987.

64

Top-loaded, Asymmetric, Shunt-tuned Low-profile Monopole:
Reflection Coefficient and Input Impedance Locus

James McLean, Heinrich Foltz
University of Texas Pan American
Edinburg, TX
tel. (512) 450-0544
fax. (512) 450-0415
e-mail: mclean@ccsi.com

Data measured in situ
HP 8714 ANA
z0 = 50. Ohms
April 1, 1999

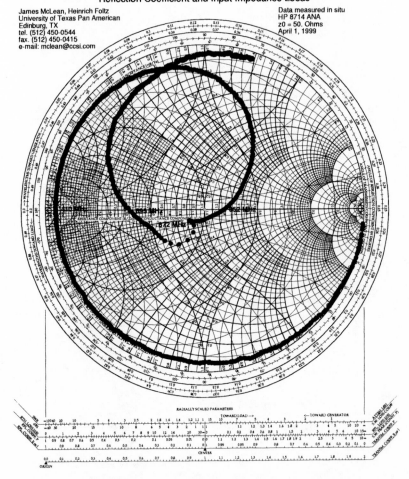

Figure 6: **Impedance Locus for Low-profile Asymmetric Monopole**

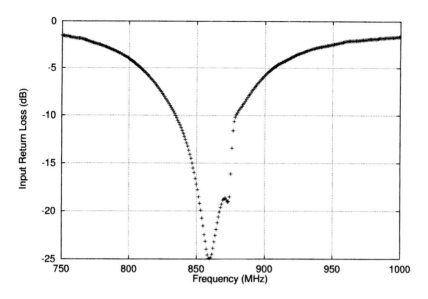

Figure 7: **Input Return Loss for Low-profile Asymmetric Monopole**

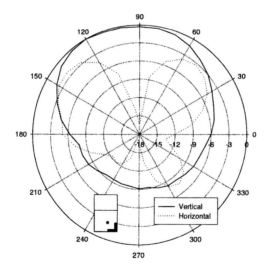

Figure 8: **Azimuthal Pattern Plot for Low-profile Asymmetric Monopole**

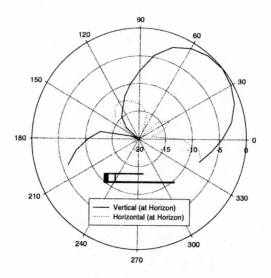

Figure 9: **Elevation Pattern Plot for Low-profile Asymmetric Monopole**

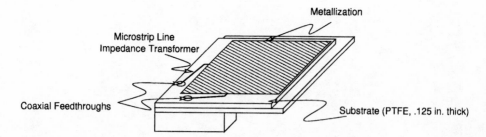

Figure 10: **Dual-polarization Microstrip Patch Antenna with Transmission Line Transformers**

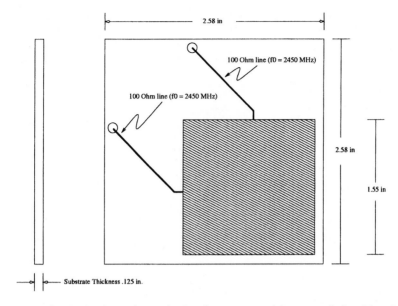

Figure 11: **Dual-polarization Microstrip Patch Antenna with Transmission Line Transformers**

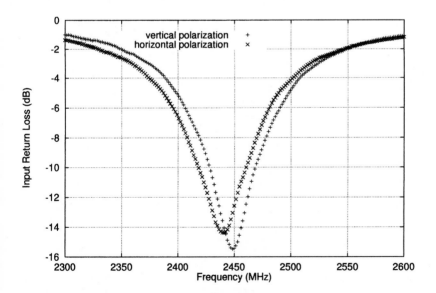

Figure 12: **Input Return Loss for Dual-polarization 2450 MHZ ISM Band Microstrip Patch Antenna with Transmission Line Transformers**

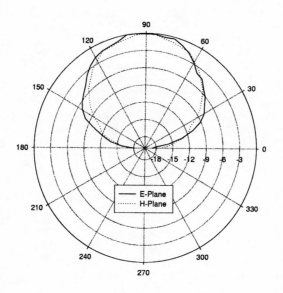

Figure 13: **E and H plane patterns for 2450 MHZ ISM Band Microstrip Patch Antenna**

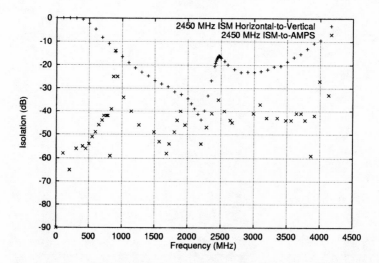

Figure 14: **Measured Isolation Data: (+) 2450 MHz ISM Horizontal Polarization Input Port to 2450 MHz ISM Vertical Polarization Input Port and (x) 2450 MHz Vertical Polarization Input Port to AMPS (Low-profile Asymmetric Monopole) Input Port**

Self-Calibration Scheme for Antenna Arrays Using the Combined Array Signal

Mark Wiegmann
University of Paderborn
Dept. of Communications Engineering
D-33095 Paderborn Germany
email: wiegmann@nt.uni-paderborn.de

It is known that antenna arrays need calibration in practice. In the following, a calibration scheme for antenna arrays which employ an analogue beamforming network and a single receiver is presented. Those Arrays supply the continuously summed signals of all branches. Consequently, vector-based calibration algorithms which depend upon the complete array response vector cannot be employed. Therefore, the presented self-calibration scheme is developed to calibrate antenna arrays which provide only the combined array signal as a calibration quality measure. Simulations showed good performance of the calibration algorithms.

1 Motivation

Steerable or even adaptive antenna arrays offer superior performance over conventional antenna concepts in the case of fluctuating signal situations. But in practice antenna arrays need to be calibrated in order to offer their full advantages. Although research focuses more and more on arrays using digital beamforming [1][2], there are still applications for which digital beamforming is too complicated or not cost efficient. In those areas, arrays using analogue beamforming are still the best choice to make. The problem with those arrays is that only a single receiver is usually implemented which means only the combined array signal is available as a calibration quality measure. Moreover, due to the analogue beamforming it is not always possible to completely switch off single branches before summation. As a consequence, the calibration scheme presented here is developed to work with the continuously summed signals from all branches of the array.

2 The error model

A uniform linear array with N elements is considered. The errors which can be recovered by the calibration scheme are random phase and amplitude deviations in each branch. The undisturbed array output signal $y(t)$ is calculated by multiplying the complex weight vector $\mathbf{w}$ with the branch signal vector $\mathbf{x}(t)$,

$$y(t) = \mathbf{w}^{\mathbf{T}} \cdot \mathbf{x}(t) = \left(\begin{array}{ccc} w_1 & \ldots & w_N \end{array} \right) \cdot \left(\begin{array}{c} x_1(t) \\ \vdots \\ x_N(t) \end{array} \right). \tag{1}$$

The amplitude and phase deviations can now be modeled by complex factors d_n in each branch. This leads to a mathematical description with a matrix $\mathbf{D}$ which entries are all zeros except for the ones on the main diagonal. The diagonal entries $d_{nn} = d_n$, $n \in \{1, \ldots, N\}$, describe the deviations in a complex notation as discussed before. The disturbed output signal

$$\tilde{y}(t) = \mathbf{w}^{\mathbf{T}} \cdot \left(\begin{array}{c} d_1 x_1(t) \\ \vdots \\ d_N x_N(t) \end{array} \right) = \mathbf{w}^{\mathbf{T}} \mathbf{D} \mathbf{x}(t) \tag{2}$$

can be regarded as an output signal generated by a disturbed weight vector $\tilde{\mathbf{w}}$,

$$\tilde{y}(t) = \mathbf{w}^{\mathbf{T}} \mathbf{D} \mathbf{x}(t) = \left(\begin{array}{ccc} w_1 d_1 & \ldots & w_N d_N \end{array} \right) \cdot \mathbf{x}(t) = \tilde{\mathbf{w}}^{\mathbf{T}}(t) \cdot \mathbf{x}(t). \tag{3}$$

Consequently, the array can be calibrated by computing a biased weight vector

$$\mathbf{w}_{\text{cal}}^{\mathbf{T}} = \left(\begin{array}{ccc} \frac{w_1}{d_1} & \ldots & \frac{w_N}{d_N} \end{array} \right). \tag{4}$$

Employing this calibrated weight vector then allows to compensate the deviations in the array branches in combination with beamforming,

$$y(t) = \mathbf{w}_{\text{cal}}^{\mathbf{T}} \mathbf{D} \mathbf{x}(t). \tag{5}$$

Usually the deviations modeled by $\mathbf{D}$ are time-variant, such that $\mathbf{D} = \mathbf{D}(t)$. But during calibration scheme execution, these deviations and thereby $\mathbf{D}$ must remain constant. In other words, the variation of the deviations in the array must be slow compared to the calibration speed. Actually, the time variance of $\mathbf{D}$ is the reason for equipping antenna arrays with a calibration device. If $\mathbf{D}$ was constant for all times, an initial calibration after array setup would be sufficient.

Depending on the physical reasons for the deviations (e.g. mutual coupling), the latter can also be direction dependent. This means $\mathbf{D} = \mathbf{D}(\theta)$, where θ specifies the angle to which the array is steered. With this simple model the effects of mutual coupling cannot be completely compensated. But since the effects of mutual coupling cannot even be modeled correctly by more complicated matrix operations [1], the assumption of direction dependent deviations is an approximation which is not optimal but easy to handle. The consequences of this direction dependency will be considered later on, where the feeding mechanism of the calibration signal will be discussed.

3 The antenna concept

As mentioned before, an N-element uniform linear antenna array equipped with an analogue beam-forming and power combining network is considered. Figure 2 shows the possible signal flow for the

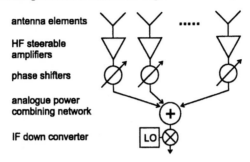

antenna elements

HF steerable
amplifiers

phase shifters

analogue power
combining network

IF down converter

Figure 1: Antenna array with analogue beamforming

implementation of a calibration device. A calibration source which emits the calibration signal is part of the calibration equipment. The form of the calibration signal depends on the environment of the antenna and can even be a continuous wave signal in the simplest case. The calibration signal has to be coupled into the antenna. After passing the antenna, the calibration signal has to be coupled out and must be fed into a measuring unit. This measuring unit passes information to the antenna processor. This may be the same processor which is already implemented for beamforming.

The coupling of the calibration signal into the antenna can be realized in different ways. One possible solution is the insertion of directional couplers into the signal path right after the antenna

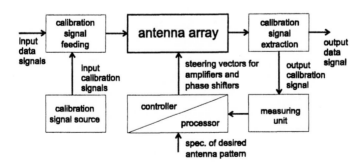

Figure 2: Signal flow in antenna array with calibration equipment

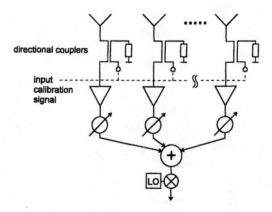

Figure 3: Calibration signal feeding through directional couplers

elements, as depicted in Figure 3. Choosing this implementation the antenna elements are excluded from the calibration system.

One of the disadvantages of this configuration is that effects as mutual coupling or array deformation cannot be calibrated this way. The exclusion of the deviations due to the anntenna aperture may seem a severe drawback, but it has to be considered that those deviations are usually time-invariant for most antennas and can be considered by an initial calibration. On the other hand, the deviations in the array branches, which are the ones covered by this realization, are in nearly all cases direction independent and thus can be calibrated with this setup in every situation. This makes the calibration signal feeding using couplers suitable for the — at least partial — calibration of mobile antennas.

A second possible way to feed the calibration signal into the antenna is the use of calibration beacons of known position in the far field of the antenna, as depicted in Figure 4. Here, the antenna elements are included in the calibration path. This leads to a direction dependency of the deviations and makes a calibration — this time for the whole antenna — also direction dependent, which means that the calibration is only valid for the direction of the received calibration signal. Therefore, the relative position of the calibration beacon to the antenna during calibration scheme execution must remain constant.

The two scenarios depicted before are quite different concerning the antenna environment and the calibrated antenna deviations. So it might be reasonable to combine both ways of calibration signal insertion. For the calibration algorithm described in this paper it is sufficient to provide a nominally coherent calibration signal in all branches of the array. For the case of feeding by a beacon, this

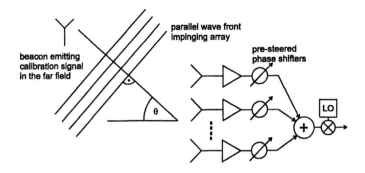

Figure 4: Calibration signal feeding by beacon in the far field

means the array is pre-steered to the known direction.

After passing the antenna, the calibration signal must be coupled out and delivered to a measuring unit. This unit must simply determine the signal power or amplitude of the calibration signal. It is not necessary to supply any phase information through the measuring unit. The information on the signal power or amplitude must be passed to the array processor. Using this information, the processor computes a new weight vector by running the calibration algorithm discussed later on.

4 The calibration algorithm

For the explanation of the calibration algorithm, a pointer representation for the signals in the array branches will be used [3]. The length of a pointer represents the signal amplitude. The orientation of a pointer in the complex plane represents the signal phase. An antenna array with a coherently fed calibration signal and a steering vector based on the N-element uniform vector

$$\mathbf{w}_{\mathrm{uniform}}^{\mathbf{T}} = \begin{pmatrix} 1 & \ldots & 1 \end{pmatrix}, \tag{6}$$

is considered. Without any deviations in the array, the summation of the branch signals — uniform pointers with uniform orientation — leads to a vector with maximum length. If the calibration signal is fed by a beacon in the far field, the array must be pre-steered to the beacon's direction, i.e. the steering vector has to have a Vandermonde-structure:

$$\mathbf{w}_{\mathrm{pre}} = \begin{pmatrix} 1 & e^{ju} & \ldots & e^{j(N-1)u} \end{pmatrix}. \tag{7}$$

The variable u depends on the parameters wavelength λ, element spacing d and signal direction θ;

$$u = 2\pi \frac{d}{\lambda} \sin \theta. \tag{8}$$

With the pre-steering and without deviations in the array, a vector with maximum length is also formed after summation. Introducing deviations in the array branches the sum vector will be shorter and will have a slightly different direction, see Figure 5.

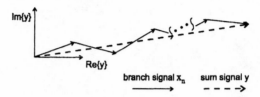

Figure 5: Pointer representation of branch signals

The calibration algorithm is divided into two parts. First, a phase calibration algorithm is executed. The amplitude calibration algorithm is to be executed afterwards as a second step.

4.1 Phase calibration

Initially, all phase shifters are set to a state, so that all branch signals are summed with the same nominal phase. This means the phase shifters are pre-steered in case of calibration signal insertion by a beacon. Or, considering a coherent feeding with couplers, the phases shifters are all set to the same state, e.g. $0°$. This state of the phase shifters is considered as the neutral state.

Due to the deviations in the array, the branch signals coming from the phase shifters do not all have the same phase. This leads to a rotated and shortened sum vector as discussed before. As a first step, the sum vector of all branch signals except the first branch signal is considered. This sum vector differs only slightly from the sum vector of all branch signals. Especially the phase difference between those two vectors is to be expected small, if there are approximately ten or more elements in the array. Then the phase shifter in the first branch is switched to a state which differs $180°$ from its initial state. Looking at the case with coherent feeding by couplers the phase shifter would be set from $0°$ to $180°$. This nominal $180°$ position of the phase shifter is considered as the reverse state. Starting with this reverse setting for the first phase shifter, its setting is altered again until the sum vector of all branch signals — including the first branch — reaches a minimum length. Practically, this means the power detected in the measuring unit becomes a minimum. This is illustrated in Figure 6. The difference between the reverse setting and the final setting with minimum length sum vector is to be recorded as the deviation to be calibrated in the first branch. After that, the phase shifter is reset to the neutral state. The procedure described for the phase shifter in the first branch

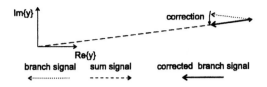

Figure 6: Pointer representation of phase calibration step

of the array has to be repeated sequentially with all the phase shifters in the other array branches. After recording the deviation for each branch in the array, one iteration of the calibration algorithm is finished.

Previous to the next iteration, each phase shifter is biased by the amount that was computed in the iteration before. This biased setting is the new neutral state. With all the phase shifters in neutral state, the vector for the sum signal should be already a little longer than it was before the last iteration. The reason is, the phase deviations in the branches are already partially calibrated.

The next iteration will be started with the new neutral state for each phase shifter. The iteration steps are to be repeated until there are no further changes determined for the phase shifters. This means, the neutral state remains unchanged after any further iteration step, because the algorithm found the optimum bias for each phase shifter. This is a reasonable stopping criterion. Especially, if the phase shifters have quantized steering ranges, it may happen that the calibration algorithm switches back and forth between two steering vectors. In this case, one calibration is as good as the other and one of the two can finally be chosen.

4.2 Amplitude calibration

After the calibration of the phase deviations in the array branches, the amplitude deviations can be calibrated. There is still the constraint that only the sum signal, i.e. the sum vector, can be measured. A result of the phase calibration is that all pointers — representing the branch signals — are parallel. A coherent calibration signal feeding or a far-field-feeding with pre-steering is assumed as before. So, the goal of the amplitude calibration is to equalize the length of the pointers.

Considering an N-element array, the amplitude calibration algorithm starts with computing the mean m of the sum signals s_n, $n \in \{1, 2, \ldots, N\}$. The sum signal s_n is determined as follows: All branch signals are summed with uniform phase. This can be done by using coherent calibration signal feeding and by setting the phase shifters to their neutral position, e.g. . The latter is the one

76

which was derived from the last step of the phase calibration. The gain of all amplifiers is set to the same nominal value, which is preferably a medium gain. This state of the amplifiers is considered as the initial neutral state. Then the phase shifter in the nth branch is set to reverse state. The resulting sum signal is denoted by s_n. Executing the described procedure for all $n \in \{1, 2, \ldots, N\}$ supplies the remaining s_n. Finally, the mean can be computed as

$$m = \frac{1}{N} \sum_{n=1}^{N} s_n. \tag{9}$$

Figure 7 shows an example with $N = 4$. After processing the mean value m, the calibration of the

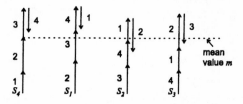

Figure 7: Pointer representation of amplitude calibration step

amplifier in the first branch can be started. The phase shifter in the first branch is set to reverse state while all the other phase shifters remain in neutral state, as explained before. The measuring unit receives the sum signal s_1. Then, the amplification of the first amplifier is set to its minimum value. The amplification is increased again and it is checked if the value of the sum signal matches the predetermined value of m. If the matching is successful, the current state of the first amplifier, i.e. its gain, is stored as its new calibrated neutral state. If the matching is impossible with any setting of the amplifier, the neutral state of the amplifier remains unchanged.

The medium value m is computed after the calibration of each branch, which also improves the speed of convergence. The amplitude calibration stops if the maximum difference between any of the sum signals s_n and the medium value m remains under an appropriately chosen boundary η. If the maximum difference exceeds this boundary, the calibration will be continued with the next branch in the array. After calibrating the last branch in the array, the calibration restarts with the first branch.

The choice of the boundary η depends on the size of the quantization steps for the gain settings of the amplifiers. If the boundary is badly chosen, the stopping criterion does not work. So the calibration should be stopped after a finite number of iterations. In this case, the states of the amplifiers have to be used for calibration for which the maximum difference between any of the sum signals s_n and the medium value m became a minimum.

This calibration algorithm does not calibrate the gain to match an exact value but — as mentioned at the beginning of the section — to match the gain to a uniform value in the array. So it can happen that the gain of the amplifiers is constantly increasing (or constantly decreasing) with every new iteration step. Finally, this would lead to a clipping of the calibration signal. To avoid this, the setting of all amplifiers should be scaled after each iteration step. This means, if the smallest gain of any of the amplifier gains exceeds an upper boundary, the gain settings of all amplifiers will be scaled down by the same factor. The other way round, if the largest gain of any of the amplifier gains falls short of a lower boundary, the gain settings of all amplifiers will be scaled up by the same factor.

5 Simulations

With the two calibration algorithms for phase and amplitude deviations, a number of simulations were carried out. Each of the entries d_n of the Matrix $\mathbf{D}$ is modeled as

$$d_n = (1 + \Delta\psi_n) \cdot \exp\left(j\Delta\varphi_n\pi\right). \tag{10}$$

The $\Delta\psi_n$ and the $\Delta\varphi_n$ are randomly chosen for each n as follows:

$$\Delta\psi_n \in [-\Delta\psi, +\Delta\psi], \Delta\psi \leq 1 \text{ and } \Delta\varphi_n \in [-\Delta\varphi, +\Delta\varphi], \Delta\varphi \leq 1 \quad \forall n. \tag{11}$$

This model allows gain deviations from complete branch failure up to doubled gain. The phase deviations can go up to $\pm\pi$ which results in completely random phases.

The following figures show simulations of calibrations with a 16-element antenna array. The range value for the phase deviations $\Delta\varphi$ was set to 0.75 and the range value for the amplitude deviations was also set to 0.75. So, severe deviations in the array branches for the phase and for the amplitude are to be expected. To visualize the effects of deviations and calibration, a uniform array factor is plotted over the equivalent direction variable u. The array factor is always normalized to its maximum. So the array factors in the plots yield 0 dB as a maximum.

The dotted line marks the original array factor without any deviations in the branches. The dashed line shows the array factor with random deviations in the branches for phase and amplitude. The solid curve shows the array factor after calibrating the deviations with the presented algorithms.

In Figure 8, the settings of the phase shifters and of the steerable amplifiers are quantized with a resolution of 8 bit. It can be seen that the algorithms are able to calibrate the array with good

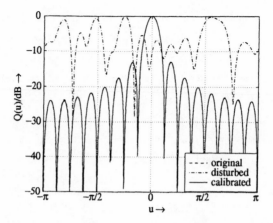

Figure 8: Calibration with 8 bit resolution for amplifiers and phase shifters

precision. The calibrated array factor and the original array factor are nearly identical. In Figure 9, the resolution for amplifiers and phase shifters was reduced to 4 bit. As a consequence, the algorithms cannot completely calibrate the deviations due to the coarse quantization. It has to be pointed out that this shortcoming is only due to the unsatisfactory resolution of the settings of the amplifiers and phase shifters. A calibration cannot be better than this with such restrictions for the steering precision.

In Figure 10 the resolution for the phase shifter is increased again to 8 bit and the resolution for

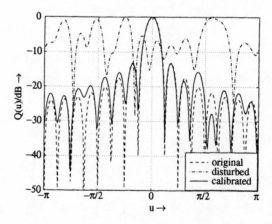

Figure 9: Calibration with 4 bit resolution for amplifiers and phase shifters

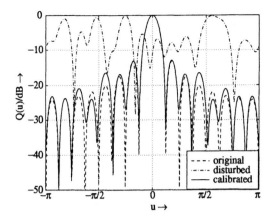

Figure 10: Calibration with 4 bit amplifier and 8 bit phase shifter resolution

the amplifiers is left at 4 bit. There, the phase deviations can be properly calibrated, which yields a symmetric array factor. The remaining difference between the calibrated array factor and the original array factor is caused by the unsatisfactory calibration of the amplitude deviations.

Figure 11 shows the situation with a resolution of 4 bit for the phase shifters and 8 bit for the amplifiers. The calibration of the phase deviations is unsatisfactory in this case and leaves a asymmetric array factor. The good calibration of the amplitudes afterwards shows only little effect.

This demonstrates that phase deviations are more critical than amplitude deviations for array

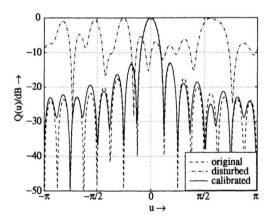

Figure 11: Calibration with 8 bit amplifier and 4 bit phase shifter resolution

processing. So, the main effort should be put into a proper calibration of the phase deviations. Another consequence is that equipping an existing array with calibration equipment and leaving the beamforming equipment unchanged may not deliver the desired results. The resolution of the beamforming devices has also to be adjusted to the expected precision. But adding a calibration system to an antenna array — as suggested here — can significantly improve performance of any antenna up to its maximum.

Moreover, the simulations showed good performance for small arrays down to element numbers of three. For large arrays — element numbers higher than thirty —, the requirements for the precision of the measuring unit rise. This is due to the small relative difference between the sum vector with all elements in neutral state on the one hand and the sum vector with a single element in reverse state on the other hand. This small relative difference might be difficult to measure practically.

6 Conclusion

A calibration scheme for antenna arrays using analogue beamforming was developed. Suggestions were made for the necessary hardware to be implemented with a calibration system. Most important, an algorithm for amplitude and phase calibration was presented. One of the important features of the algorithm is that only the continuously summed signal of all array branches is needed as the calibration quality measure. The simulations showed good performance of the calibration algorithm for various array sizes and even for severe phase and amplitude deviations in the array branches.

7 Acknowledgement

The author thanks Robert Bosch GmbH for the support and the good cooperation.

References

[1]   C.G. Brown, J.H. McClellan, E.J. Holder: "A Phase Array Calibration Technique Using Eigenstructure Methods", IEEE International Radar Conference, 1990, pp. 304-308

[2]   J. Herd: "Experimental Results from a Self-Calibrating Digital Beamforming Array", IEEE Antennas and Propagation International Symposium Digest, 1990, pp. 384-387

[3]   H.W. Kummer: "Basic Array Theory", Proceedings of the IEEE, Vol.80, No.1, January 1992, pp. 127-139

[4]   M.P. Wylie, S. Roy, R.F. Schmitt: "Self-Calibration of Linear Equi-Spaced (LES) Arrays", IEEE Proceedings ICASSP 1993, pp. I281-I283

Switched Beam Adaptive Antenna Demonstrator for UMTS Data Rates

Heinz Novak

Institut für Nachrichtentechnik und Hochfrequenztechnik (INTHFT)
Technische Universität Wien,
Gusshausstrasse 25/389, A-1040 Vienna, AUSTRIA
heinz_novak@ieee.org

Abstract:

Adaptive antennas are a promising way to increase capacity in today's mobile communication systems and will be an optional or mandatory component for the next generation. We developed a test system with a fixed beam-grid antenna to investigate the benefits that can be expected by employing such a system. The beamforming network was realized as Butler matrix in a novel single-layer structure, which allows easy and low cost manufacturing. The base station transceiver is equipped with powerful measurement equipment for online bit error measurement and for logging of the complex baseband signal with eight-fold oversampling for post processing. This system allows thorough evaluation of switching algorithms, gives insight into error mechanisms and helps to investigate channel properties.

Introduction

In today's mobile communication systems radio frequency spectrum is a limited resource. To overcome this limitation considerable research is going on to use the given frequency bands as efficiently as possible. The deployment of adaptive antennas is one way to increase spectrum efficiency in mobile communication systems /1/,/2/.

Compared to today's mobile communication systems, where users are separated either in time, frequency or code domain, in adaptive antenna systems the users are additionally separated by the angle at which they are seen from the base station. With this information the base station does not need to radiate energy isotropically over the whole cell, but it can steer the antenna beam into the direction of the user. Thus energy is only radiated into a small angular range and the signal is only deployed where it is needed, with no or little signal energy being radiated into the rest of the cell. The previous also holds for reception.

Pointing the beam to the user leads to an increased gain into the user's direction and to an improvement of the signal-to-noise ratio of the received signal. At the same time the reduced gain in all other directions will help to reduce interference from other users and to suppress multipath components from the same user, which will result in a lowered delay spread. All these effects lead to a capacity increase.

To enable the change of the antenna pattern, adaptive antennas are usually built as antenna arrays. The single elements are fed by replicas of the same signal, which are phase shifted and attenuated, to form an antenna pattern. This signal conditioning can be done at RF or in base band (BB), which leads to the alternatives of RF and BB beamforming. Another classification of adaptive antennas distinguishes between fixed beam grid antennas, which offer a fixed number of beams, and fully adaptive arrays, which allow to point the main beam and the nulls into arbitrary directions.

In this paper an adaptive antenna system with RF beamforming and a fixed beam grid is presented. On the one hand this is not optimum, since it reduces the degrees of freedom, but on the other hand the complexity and cost of the antenna system can be reduced significantly. A further benefit of switched beam is that the output is at RF level, which makes upgrading of an existing system very easy.

Switched Beam Testbed (SITE) Hardware

System Concept

The Switched Beam Testbed SITE is shown in Fig. 1. It consists of the base station with the adaptive antenna, two mobile stations and a control PC, which acts as the man-machine-interface.

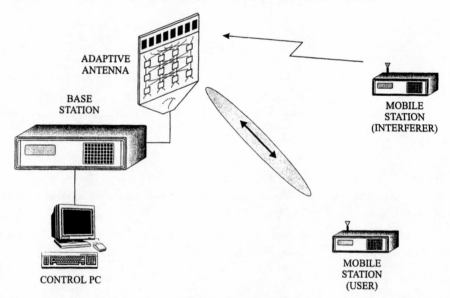

Figure 1 The Switched Beam Testbed (SITE) system configuration

The adaptive antenna, which allows to select one of a fixed number of beams, is connected to the base station, which directs the beam to the desired user. This can be done by using a received signal strength indicator (RSSI) or a bit error rate (BER) based criterion. Signals are transmitted to and received from a mobile station, which uses an omni-directional antenna. The transmitted signals can be either speech data from the user or a fixed bit sequence if BER measurements are done.

To test interference immunity of the system, a second mobile station can be added to the setup to act as an undesired user. BER measurement is done in the base station by a dedicated unit, which not only does online BER calculation, but also can record the received signals for post processing. Operation of the whole system is managed by a control PC.

The air interface of the SITE system uses a time division multiple access (TDMA) / frequency division multiple access (FDMA) / time division duplex (TDD) transmission. The parameters of the air interface are summed up in Table 1.

transmission format	TDMA / FDMA / TDD
frequency range	2.4000 ... 2.4835 GHz
bit rate	1.152 Mbit/s
channel spacing	1.728 MHz
modulation	gaussian minimum shift keying (GMSK)
modulation index	BT = 0.5
frame length	10 ms
bursts per frame	24

Table 1 SITE system parameters

As can be seen the protocol and the modulation format for transmission are equal to DECT (Digital Enhanced Cordless Telecommunications), whereas the frequency range is shifted to the ISM (Industrial, Scientific and Medical) band at 2.45 GHz. The data rates of more than one Mbit/s make also valid for third generation systems, e.g. UMTS (Universal Mobile Telecommunications System).

Building Blocks

Adaptive antenna

The adaptive antenna consists of the antenna array, the beamforming network and a selection switch /3/. The antenna array uses 8 active and 2 dummy patch antenna elements. It is implemented in a Strip - Slot - Foam - Inverted Patch (SSFIP) structure /4/. This microstrip patch antenna design results in high bandwidth compared to standard microstrip antennas with increased efficiency, low weight and low cost. In the SSFIP structure the antenna is built on a thick substrate of low-permittivity material in order to get high bandwidth. The metal patches are covered by a thin plastic layer mounted in such a way that the printed patch is directly positioned on top of the foam (inverted patch technique). Thus you get both protection from the environment and a material to mount the patches on. The plastic layer is so thin that its effects on the antenna characteristics are negligible. To minimize mutual coupling between the patches a special feeding technique - aperture coupling - is used. Extensive simulations showed that a H-

84

shaped slot gives minimum mutual coupling of less than -15 dB, while still guaranteeing high efficiency. The dummy antenna elements, one on each side, render the effects of mutual coupling for all elements.

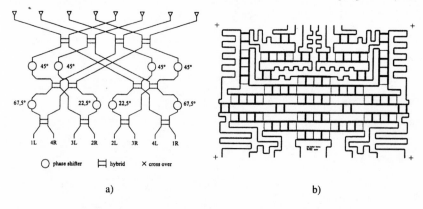

a)   b)

Figure 2 Butler matrix layout: a) logical structure, b) final layout

The active elements are connected to the outputs of a 8x8 single layer Butler matrix, which performs the beam-forming. The Butler matrix, as shown in Fig. 2, typically is a N input and N output network consisting of phase shifters, hybrids and crossovers. It was built in microstrip technique as single layer structure, which avoids the need for the signal to change from one layer to an other and thus makes production very cheap. In such an implementation the crossovers are realized as two cascaded hybrids/5/.

A special selector switch allows to choose either a single beam or a combination of two neighbouring beams for transmission and reception. The combination technique substantially reduces the side lobes of the antenna pattern by 6 dB as compared to the single beam approach.

Base station

The base station incorporates a complete RF transceiver, a base band processor, a microcontroller, which communicates with the PC and sets all parameters of the transceiver, and a powerful error measurement unit.

Fig. 3 shows the segmentation of the SITE base station, which is not a purely logical one, but reflects the physical setup. This means that each of the boxes in the figure is a plug-in unit mounted in a 19" system case and conneted to the backplane system bus of the system case.

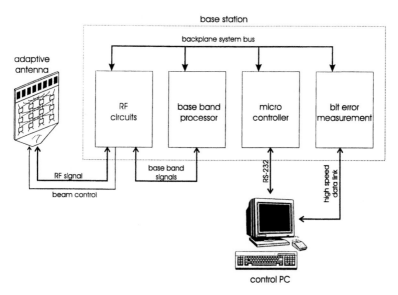

Figure 3 SITE base station system

In the *RF unit* all RF circuits, including the synthesizer, the up and down converters, the AGC circuits and RSSI measurement, are combined. The heterodyne transceiver converts the incoming signal to an intermediate frequency of 110.592 MHz and from there to quadrature base band for reception and vice versa for transmission. For level control of the incoming signal an open loop automatic gain control (AGC) is used, which yields a reduced dynamic range of the baseband signal by 40 dB as compared to the RF signal. A dual synthesizer produces the two low phase noise local oscillator (LO) signals. The transmit and receive signals are exchanged between the RF and base band units in complex base band format on the front panel. The control signals are distributed between all units over the backplane system bus.

In the *base band processor* the quadrature base band signals are demodulated, and bit and slot synchronization is performed. For transmission, either a speech signal or a pseudo random signal can be used. Synchronization information is added to the signal before the modulated base band quadrature signal is output to the transceiver.

The connection of the *micro controller* to the PC is done via a RS-232 link, which offers connectivity to all standard PCs but does not allow very high data transfer rates. When the quadrature signals are recorded, the 420 bit in one time slot amount to 6720 byte of data per time slot (420 bit * 8 fold oversampling * 8 bit I and Q). If one time-slot is transmitted per frame, the data rate becomes 672 kbyte/s (frame duration = 10ms), which is much more then the serial connection can handle.

In the *bit error measurement (BER) unit* the demodulated signals are compared to the known reference sequence and the resulting bit error rate is calculated. For post processing it is possible to record the received and analog-to-digital converted quadrature signals at an oversampling rate of eight, the demodulated and synchronized data bits and the demodulated data bits without synchronization at an oversampling rate of eight. To transfer the quadrature data to the PC a dedicated high speed link was implemented, which connects the BER unit to the PC at a data rate of above 1 Mbyte/s.

Mobile stations and control PC

The mobile stations are equipped with omni-directional antennas. For simplicity of operation and design, the mobile station transceivers are identical to the base station ones. No error measurement is done in the mobile station. The PC controls the parameters of the mobile station either via a RS232 cable connected to the mobile, by a simple control unit attached to the mobile or by radio commands via the base station.

The whole system can be controlled from the PC, where MS Windows based software offers easy access to all parameters of the system.

Measurement Options and Modes of Operation

Switching Algorithms

When thinking about a switched beam system, the method of selecting the used beam is the main issue. A straight forward way of classifying the received signals on the available beams is according to their amplitude or signal strength, thus called received signal strength indicator (RSSI) based. This method is very simple, but it does not take into consideration that the strongest signal will not necessarily have the lowest bit error rate. Thus another method bases the selection on bit error rate (BER), which is measured in all beams. This method provides a much more relevant criterion, but to measure the bit error rate in several beams, it is necessary to introduce a dedicated measurement burst in the transmission and by this reducing the over all system capacity. Both methods are implemented in SITE, to allow quantification of their difference in performance.

In the *RSSI based algorithm* the antenna beam is switched from the first to the last possible beam position and the RSSI value is measured for all beams, as shown in figure 4. This is done during the first 32 bits of the burst, which contains the synchronization field and the start flag. Both can not be used for synchronization and we have to assume that the synchronization gained in the last frame is still valid. This leaves the receiver the duration of four bits or 3.48 µs to measure and sample the RSSI value for each beam. From this it can be seen that a very fast RSSI circuit has to be used if the RSSI based algorithm is used. Immediately after all beams have been measured, the beam with the biggest RSSI value is selected for the reception of the remaining part of the burst.

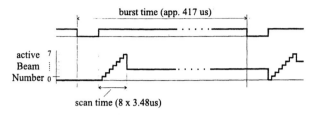

Figure 4 Received signal strength indicator (RSSI) based switching

The *BER based selection* is shown in figure 5. In this mode a whole burst, the measurement burst, is reserved to assess the quality of the signals received with the different beams. If we assume that we want to use the 32 bit synchronization field to synchronize the measurement burst itself, we have 388 of the 420 bits left for bit error measurement in the 8 beams. Thus 48 bits are used for each beam and a minimum bit rate of 0.021 can be detected. This criterion gives us a much sounder bases to select the best beam, but we pay for this with the loss of a complete burst. So the number of burst that can be used for data transmission is reduced from 24 to 23, but as we always use pairs of bursts, we loose one of the 12 available burst pairs, or 8.3 percent of the capacity. This amount must be considered, when capacity gains are calculated from improvements in BER or SIR. The data burst is transmitted and received on the beam with the smallest bit error rate.

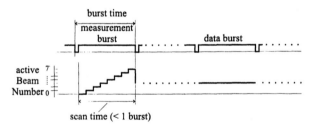

Figure 5 Bit error rate (BER) based switching

Measurement options

Bit error measurement: The main measurements, which will be done in the testbed, are bit error measurements. The mobile station transmits a fixed bit sequence, which is received by the base station and compared to the known sequence. The number of bits, which are counted for the calculation of the error percentage, can be varied to allow different levels of accuracy. For accurate measurements the number of measured bits must be high and thus the time between new values of bit error rate will also be high. In a modified version of the measurement, the bit sequence is transmitted from the base station, received by the mobile station and transmitted back to the base station. This measurement includes errors on the uplink and the downlink and thus reflects a mixture of errors in both directions. If we assume that double errors, which means that a bit has an error in the up- and downlink and thus seems correct, can be neglected and that up- and downlink contribute equally to the errors, we can calculate the errors in one direction by simply dividing the result by two. This assumption still has to be verified by measurements.

Data logging: It is not always sufficient to know the percentage of the erroneously received bits. To study error mechanisms it is necessary to investigate the temporal order of the errors. To support this it is possible not only to record the temporal succession of the errors, but also to record the raw data and the base band quadrature signals. The raw data is a binary data stream at an oversampling rate of eight, which is the output signal of the demodulator, whereas the quadrature base band signal is the eight bit representation of the analog signal also sampled at an oversampling rate of eight. Those signals allow extensive data analysis in a post processing step. Thus we can evaluate the influence of different sampling times or different demodulators on the bit error rate.

Receive level: During the on-line bit error measurement the level of the received signal is permanently monitored and is shown on the user interface together with the measured bit error rate.

Antenna Setup: The adaptive antenna array offers two different configurations. The first one uses eight beams, which feature rather high sidelobes, whereas the second one offers seven different beams, which are slightly wider, but have much lower sidelobes. This second configuration is obtained by combining two neighbouring beams, which leads to a tapered illumination of the array and thus to lower side lobes. In both configurations the selection of the beam, which is used for transmission, can be done by the user on the PC or automatically by the system according to one of the selection methods described in the previous chapter.

Measurements

Fig. 6 shows a measured antenna diagram for the 7-beam configuration of the Butler matrix. In this configuration a side lobe suppression of 17 dB can be reached.

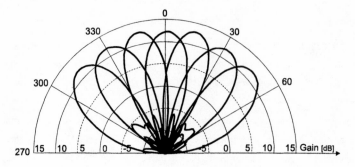

Figure 6 Measured antenna diagram

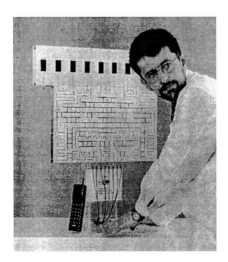

Figure 7 Adaptive antenna with Butler matrix

Conclusions

We have implemented a switched beam adaptive antenna system with powerful measurement capabilities to investigate the benefits achievable with fixed beam grid antennas. By implementing the Butler matrix beam forming network in a novel single layer structure, the production process is extremely low cost. A special combining technique reduces the sidelobes of the antenna beams to -17 dB below the main beam. The system supports data rates in excess of one Mbit/s and thus will make the results also valid for third generation systems. The amount of inter-symbol interference reduction with this antenna type will be studied in the following months. The recorded complex base band signals give insight into error mechanisms. All modules of the demonstrator and the necessary mobile units, including various patch antenna arrays, have been developed and implemented at our institute.

Acknowledgment: I thank Prof. Ernst Bonek for extensive discussions and his support and encouragement.

References

[1] Winters, J. H., Smart Antennas for Wireless Systems, IEEE Personal Communications Magazine, February 1998.

[2] Ho, M., Stüber, G.L., Austin, M.D., Performance of Switched-Beam Smart Antennas for Cellular Radio Systems, IEEE Trans. Veh. Technol., Vol. 47, No. 1, Feb. 1998

[3] Novak, H. A single-layer 8x8 Butler matrix with patch antenna, MTT-S European Wireless '98, Amsterdam, October 8-9, 1998, pp. 25-29.

[4] Zürcher, J. F., The SSFIP: A Global Concept for High-Performance Broadband Planar Antennas, Electronic Letters, EL-24(23):1433-1435, November 1988

[5] Nord, H., Implementation of a 8x8-Butler Matrix in Microstrip, Diploma Thesis, Institut für Nachrichtentechnik und Hochfrequenztechnik, Technische Universität Wien, Austria, Dec. 1997

UMTS Radio Network Simulation with Smart Antennas

Boukalov O. Adrian, Sven-Gustav Häggman
Communication laboratory, Institute of Radio Communications (IRC)
Helsinki University of Technology, P. O. Box 3000, FIN-02015 HUT, Finland.
Phone: +358 9 451 [2317, 2340] / Fax: +358 9 451 2345
E-mail: [Adrian.Boukalov, Sven-Gustav.Haggman]@hut.fi

Abstract - Smart antennas (SA) system integration into different types of cellular environments requires simulation tool which able to take into account radio propagation, network control, users' behaviour, traffic and SA algorithms simultaneously. DS-CDMA radio network simulation tool "NetSim" was further extended in order to simulate different types of smart antennas like switched beam and adaptive antennas with different types of beamforming algorithms. Study of radio network control functions, such as: admission, handover and power control together with different spatial processing algorithms became possible. Capacity improvement of CDMA network with different types of SA was studied by simulation. Traffic model of multi-bit rate services was included in the simulations.

1. Introduction

Spatial processing technology considered as a "last frontier in the battle" for cellular system capacity with limited amount of spectrum. There are number of SA commercial products already available on the market. The main advantages expected from SA technology are:
- Higher sensitive reception
- Interference cancellation in uplink and downlink
- Mitigation effects of multipath fading

On the system level, they provide higher capacity, extended range, improved coverage by "in-filling" dead spots, higher quality of services, lower power consumption at the mobile and improved power control.

Spatial and Spatial-Temporal (ST) processing in CDMA has several distinguishable features. In non-multiuser case all other users are seen as interference to each other and there are many weaker co-channel interference (CCI) at the uplink. Multipath gives rise to the multiple access interference (MAI) due to the losses of codes orthogonality. Inter symbol interference compensation has less importance in CDMA than interchip interference. Wideband beamforming realisations and methods of angle of arrival (AoA) estimation are different from narrowband. Among the proposed wideband beamformer (BF) realisations there are switched- beam approach, bearing estimation techniques [1], Eigenfilter techniques [2]. Training signals can be successfully used in wideband beamforming and minimum square error (MSE) criteria is used for weights adaptation. There are number of signal structure based beamforming methods like code-filtering approach proposed in [3] and multi-target algorithms [4] which combines information of the spreading signal and the constant modulus property in adaptation of the weight vector.

In CDMA RAKE receiver is followed by beamformer. Two dimensional 2D-RAKE receivers where MSE beamformer [5] or beamformer based on code-filtering [3] for each path is followed by conventional RAKE receiver are proposed. Space- time (ST) RAKE reduces MAI and thus improves coverage and capacity. Such a receiver structure has an additional degree of freedom and can be optimised to achieve improved coverage or capacity by reducing inter- or intracell CCI by beamforming.

Multi-user space-time maximum likelihood (MU-ST-MLSE) receiver for CDMA was proposed in [6] but practical implementation is extremely complex. This type of SA receiver has computational complexity linear to the number of users and the same degree of the near-far resistance and error rate performance as optimum MU receiver. MU-ST-MLSE requires knowledge of the all user's channels.

As it was shown in [1], sophisticated spatial-based blind methods are considerably less efficient for low SNR and it was perhaps one of the reason of more extensive research in the area of switched-beam solutions for system with IS-95 air interface during last several years. User dedicated pilots at the up- and down-links of the UMTS air interface give additional advantage for MSE methods especially in highly loaded cells. In multi-bit rate CDMA SA receiver can successfully cancel interference coming from the limited number of high bit rate users, thus considerably increase system capacity.

There are number of CDMA networks system level simulation studies with SA [7 -11]. Some of them use deterministic channel models where propagation data obtained with raytracing method [7]. All studies assume only voice service supported by the network and there are no known simulation tools which include simultaneously models of radio network control functionalities, deterministic spatial channel and smart antenna receiver.

"NetSim" was developed to study cellular networks control algorithms performance and planning strategies of the third generation cellular systems which will be able to support multi bit rate services. "NetSim" provides detailed information about system capacity, coverage and network control algorithms performances. "NetSim" output files consist of information about call dropping, blocking and temporal and spatial references of these events. Obtained statistics can be easily collected and translated into visual form with help of MATLAB or other tools. "NetSim" is written on C- language and can be updated for different radio interfaces (GSM, IS-95, UTRA/W-CDMA), various statistical and deterministic channel models and different types of radio network control algorithms. "NetSim" was successfully used in the are number of research works [12 ,13]. "NetSim" is a time driven simulator, which purpose is to fill the gap between simulators developed for study link level signal processing algorithms like COSSAP and higher level network simulators like OPNET and BONES. "NetSim" can be easily extended to include link level simulations or fixed network simulation and can be combined with other tools. "NetSim" can simulate: users behaviour, various types of teletraffic, interference, power and admission control algorithms, adaptive antennas beamforming, soft and hard handover. Current version uses a deterministic propagation model for micro-cellular urban environment based on raylaunching method. Deterministic model gives opportunity to study different radio network planning methodologies. Existing propagation model also provides information about spatial properties of radio channel for simulations radio network with adaptive antennas.

"NetSim" is a complex simulation tool and it should be always optimised according to specific task in order to avoid memory overload or/and prevent excessive simulation time.

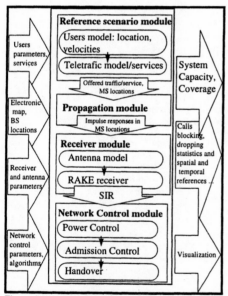

Figure 1. *"NetSim" structure.*

2. "NetSim" structure

Main elements of the "NetSim" structure are shown on the Figure 1. and they will be discussed n more detailed further. "NetSim" structure can be logically divided into four main modules related to the reference scenario, radio propagation, signal processing algorithms at the receiver and radio network control .

Reference scenario module
The output data of the reference scenario module are MS location, velocity and activated service type. Module consists of users and teletraffic models.

Users model generates users location and velocities. Pedestrian users model generate MS locations randomly distributed along predefined route. The call birth is a Poisson distributed random variable with mean birth rate per hour defined in simulation set-up. Duration of voice call is 120 seconds. The velocity of a mobile station is a Gaussian distributed random variable with mean 0 m/s and with standard deviation 1 m/s. There is also models for fixed domestic users and car passengers. Fixed and moving users models can be activated simultaneously. Spatially non-uniform users distribution modelled by Gaussian density function with different values of sigma.

Teletraffic model simulates teletraffic related to different services and control channels activity. This model simulates protocols of the physical channels and call control functionalities. It is possible to assign certain set of services to the specific users group, for example, high bit rate services and voice for the domestic users and voice- only type of service for the cars passengers. Conversation phase traffic model of is based on the so-called modified eight-state Brady model [14] which takes into consideration voice activity detection. "NetSim" has also teletraffic models of data transmission related to www browsing in Internet and video calls. Traffic model of the packet transmission will be included in the "NetSim" in the near future .

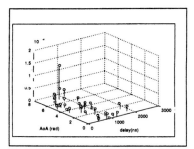

Fig.2. *Impulse response for LOS propagation*

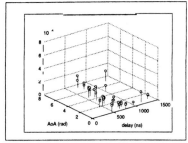

Fig. 3. *Impulse response for NLOS propagation*

Propagation module
Propagation module calculates received signal power for the each MS locations. It is based on propagation data processing obtained by the raylaunching program [15]. The raylaunching program uses electronic map for impulse response calculations. The input parameters of raylaunching program are locations of building's walls, BS position and antenna height, users locations. Propagation data are calculated for each BS and in some cases for the each antenna element. Electronic map is comprised of buildings walls whose electrical properties are known. In the current version of raylaunching program it is assumed that BS antenna height is well below rooftop level and all propagation phenomena are taking place in the street canyon. Ground reflection is considered. Because the distance between two calculated impulse responses is a quarter of wavelength, the cubic convolution interpolation method is used to predict received signal strength when mobile stations are between discrete locations.

94

The output files of raylaunching program consist of information about each path delay t_k, amplitude a_k, phase θ_k and AoA $_k$. It can be expressed analytically as

$$h(t) = \sum_{k=0}^{K-1} a_k \delta(t - t_k) \exp(j\theta_k) , \text{ where } K\text{- is the path number.}$$

Three dimensional plot of impulse responses for line- of site propagation (LOS) and non-line of site (NLOS) propagation are presented on the Figures 2 and 3.

To be able to carry out system level simulations with SA when antennas algorithm use directional information about radio channel program which extract the strongest rays in the impulse response and at the same time keeps pathloss values unchanged was developed. Thus considerable memory saving at the expense of accuracy are achieved. Propagation module can use statistical channel models or information obtained by the radio channel measurements. Simple analytic models can be incorporated in propagation module when rough simulations results are needed very fast or computer memory is a limiting factor. Three dimensional ray tracing model will be used in future for simulation network with mixed overlaying cells architecture. Simultaneous impulse response calculations and network level simulations allow to avoid interpolation and memory overload at the expense of simulation time. Parallel processing of impulse responses for each BS can considerably save program execution time in this situation.

Receiver Module

Receiver module simulates antenna, receiver algorithms and calculates average received signal power. The output information of this module is post -processing signal to interference (SIR) ratio for the each link.

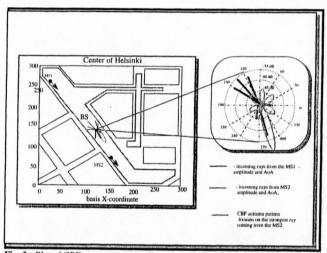

Fig. 3. *Plot of CBF antenna pattern. Two users and LOS propagation scenario*

Antenna model can simulate different antennas types at BS :
- single omnidirectional and directional antennas (sectorization);
- distributed antennas (spatial diversity);
- switched beam antennas;
- adaptive beamforming based on AoA estimation or/and reference signal;
- combined adaptive beamforming and sectorization;

Two dimensional 2D-RAKE receiver is modelled in the "NetSim". In 2D-RAKE separate beamformers are assigned to the each finger of the RAKE receiver. There are two beamformers models implemented in the "NetSim". The first model based on AoA estimation. The second one is MSE beamformer. User dedicated pilot of UMTS air interface can be used as a reference in the second model.

Conventional beamforming (CBF) chosen among beamforming methods based on AoA estimation. It provides beam steering toward direction of path with maximum power (see Fig.3.). In this model AoA estimation and tracking are assumed to be perfect. Optimisation procedure are not currently included in AoA based beamforming model. Antenna is considered as a singe element antenna with rotating pattern with fixed shape. It possible to narrow and widen beam pattern by changing amount of antenna elements. Model of AoA estimation error can be activated in simulations.

Reference signal based adaptive beamforming optimally combine incoming signals. In this case antenna model uses impulse response information obtained by raylaunching program individually for each antenna element. A number of well known adaptive algorithms like LMS and RLS are used. The *2D-RAKE receiver and interference model* calculates the signal-to-interference ratio after despreading and combining in all radio links. The total signal-to-interference ratio is obtained by summing the signal-to-interference ratios of different paths corresponding to maximal ratio combining. Another technique that can also be used is the selection combining. In this technique the receiver selects the path with the best signal-to-interference ratio. The interference consists of the thermal noise of the receiver, self-interference, and interference from the users in the same radio channel. Self-interference means here interference which is caused by the multipath propagation. All interference is modelled as additive Gaussian white noise. Received signal averaging with first-order Buttrworth filter is followed by interference model. Time constant of the filter can be selected during simulation set-up. Antenna model can be activated in the up-link only and in the both directions simultaneously. To provide information for weights adaptation at the down-link, retransmission scheme is modelled.

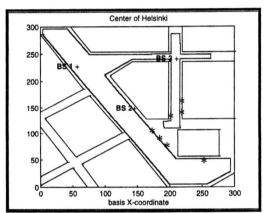

Fig. 4. *Location of the BS on the electronic map and locations(*) of the places with highest dropping probability value.*

Radio Network Control Module

Radio network control module simulates power control, admission and handover control algorithms. Network control model is also responsible for the constant SIR monitoring of all active radio links. The output data of network control module are amount of call dropping and blocking events and other system control failures with correspondent spatial and temporal references .

Power Control Model for up-link and down-link uses SIR based distributed power control algorithm [16]. Program includes return channel error and loop delay models. Outer loop power control model are included in this model to adjust SIR_{target} according to commands of radio resource management. Open loop power control is used during call initialisation period. During the initialisation phase the transmitted power of mobile station is increased till the base station receiver will detect random access signal or initialisation time will exceed specified limit. Values of SIR_{target} are largely depend on service type to be activated. Power control step size can be adaptively adjusted for each BS. During soft handover if at the any of connected links will originate command "down" power control obey it and decrease transmitted power, otherwise, it increase power. As a result there is at least one BS to provide coverage.

Admission Control Model in the current "NetSim" version uses information about total received power at the base station, signal quality measurements and their variation in time for making decision about admission. In the simulation centralised admission control is used. It means that average received power of all base stations is available for the admission control. Decision related to the new user admission is based on the maximum allowable received power in the cell (cell load) and link quality measurements during limited admission time.

Handover Model simulate soft and hard handover algorithms. Mobile assisted (MAHO) soft handover used in the current version of "NetSim". It is possible to select macrodiversity order which is the number of BS involved in the handover process, macrodiversity margin (the difference between the strongest and weakest BS in the active set). Among the adjustable handover parameters are the add- and drop- times and the hysteresys which define the difference between the weakest BS in the active set and the strongest BS outside the active set to be placed instead the weakest one. More complex centralised power control, BS assignment algorithms can be included in the network control module. Hard handover will be used for packet transmission simulations.

Analysis of the results

"NetSim" collects information and time related references of the calls dropping, blocking and admission rejection events. Erlang capacity can be defined as a number of users which can be admitted to the system when the blocking probability does not exceed 0.02. Obtained by simulation statistics allow to obtain outage probability as the probability of the event that SIR of the ongoing call falls below required level for that service. A commonly accepted values which determine the capacity of DS-CDMA system is 1% outage. Spectrum efficiency also can be easily calculated based on output data obtained with "NetSim". Coverage aspects of network planning can be studied since the spatial references of these events on electronic map are available (Fig.4) .

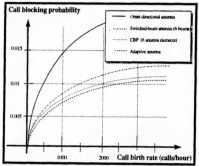

Fig. 5. *System capacity with different types of SA. Voice only service supported by the network.*

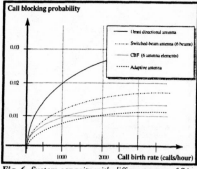

Fig. 6. *System capacity with different types of SA. Mix of voice and high bit rate services.*

3. DS-CDMA network simulations with SA

A number of DS-CDMA network simulations was done with different types of SA. Impulse responses were obtained with raylaunching program which used electronic map of Helsinki downtown. It was assumed that there are three BS with overlapping coverage in this area (Fig.4.).

In the first set of simulations pedestrian users are randomly distributed along predefined route which is located inside coverage of the network. All of the users can make only voice calls with duration 120 s. In this simulations bit rate was set-up to 16 kbits/s, processing gain was 24 dB and $SIR_{target} = 7$ dB. 2D- RAKE receiver had five fingers with one beamformer assigned for the each finger. Four antennas types were studied: omnidirectional antenna, switched beam antenna with six beams, conventional beamformer and adaptive antenna with reference signal based algorithm. Adaptive antenna include 6 elements with half wavelength spacing between them.

Simulation results on the Fig 5. show 3 times capacity improvements with six sectors switched beam antenna at the up-link and 4 and 4.7 times capacity improvements with CBF and adaptive antenna.

Second simulation includes two type of users: 90 % pedestrians with voice only type of service and 10 % of domestic users which use only calls with high bit rate. Channel bit rate for high bit rate services was set-up to 256 kbps. In simulations the same SA model was used as in the first set. Obtained results (Fig. 6) show 3.5 times improvement with switched beam antenna , 5 and 5.5 times improvements with CBF and adaptive antenna.

In simulations with switched beam antennas which had 3-4 elements it was observed a 10 -20 % capacity reduction when the beams orientation was not aliened toward street direction where largest offered traffic is expected. This effects was reduced with larger amount of beams.

4. Conclusions

Given an introduction into "NetSim" simulation tool which includes users model with different types of mobility and services. Radio network control functions such as handover, admission and power control are modelled with "NetSim". Several beamforming algorithms are incorporated into "NetSim" to work together with spatial radio channel models. Presented simulation results are based on raylaunching propagation model. Statistical propagation model and data obtained from channel measurements in the city area on the electronic map, where raylaunching results were obtained, will be used in the future simulations.

Due to complex structure and large amount of input data required for simulations, the carefully made optimisation of input data, simulation set-up and program structure are needed.

Simulation results shows substantial capacity improvements with SA technology. Threefold capacity improvement can be obtained even with simplest switch - beam antennas.

Switched beam antennas orientation can be critical in the urban areas when the amount of beams are less then 5 -6.

System aspects of SA: radio resources management with SA , dynamic behaviour of SA algorithms, link level control protocols and network planning with SA are supposed to be with "NetSim" studied in the future.

References

[1] B. Suard, *et al.*, "Performance of CDMA Mobile Communication Systems Using Antenna Array," *Proc. ICASSP*, Mineapolis, MN, Apr. 1993, pp. 1736 - 39.
[2] P.M. Grant, J.S. Thompson and B. Mulgrew, " Adaptive Arrays for Narrowband CDMA base stations", *Electronics & Communication Engineering Journal*, August 1998, pp. 156 - 166.
[3] A. F. Naguib, Adaptive Antennas for CDMA Wireless Networks", *Ph.D. Dissertation*, Stanford University, August 1996, pp. 60 -65.
[4] Z. Rong, T. S. Rapport, P. Petrus, J. H. Reed "Simulation of Multi-target Adaptive Array Algorithms for Wireless CDMA Systems", *Proc. VTC'97*, Phoenix, AZ, May 1997, pp. 1- 5.
[5] A. Naguib and A.J. Paulraj " Performance of CDMA Cellular Networks with Base Station Antenna Arrays," *Proc. Int'. Zurich Sem. Dig. Commun., Zurich, Switzerland, Mar. 1994, pp. 87 -100.*
[6] R. Kohno, "Spatial and Temporal Communication Theory Using Adaptive Antenna Array", *IEEE Personal Communications*, February 1998, pp. 28 - 35.
[7] S.C Swales *et.al.*, "The Performance Enhancement of Multibeam Adaptive Base-Station Antennas for Cellular Land Mobile Radio Systems," *IEEE Trans. on Vehicular Technology*, Vol. 39, No. 1, February 1990, pp. 56-67.

98

[8]Y. Li, M. J.Feuerstein, D. O. Reudink, "Performance Evaluation of a Cellular Base Station Multi-beam Antenna," *IEEE Trans. on Vehicular Technology*, Vol. 46, No. 1, February 1997, pp. 1-9.
[9] J. C. Liberti, T. S. Rappaport, "Analytical Results for Capacity Improvements in CDMA," *IEEE Trans. on Vehicular Technology*, Vol. 43, No. 3, August 1994, pp. 680-690.
[10] J. C. Liberti, T. S. Rappaport, "Analysis of CDMA Cellular Radio Systems Employing Adaptive Antennas in Multipath Environments, " *V TC'96*, Georgia, GA, April 29-May 1, 1996, pp. 1076-1080.
[11] A. F. Naguib, A. Paulraj, T. Kailath, "Capacity Improvement with Base-Station Antenna Arrays in Cellular CDMA," *IEEE Trans.on Vehicular Technology*, Vol. 43, No. 3, August 1994, pp. 691-698.
[12] Boukalov Adrian, Sven-Gustav Haggman, Anti Pietila "The Impact of a Non-uniform Spatial Traffic Distribution on the CDMA Cellular Network System Parameters", *ICPWC'99*, Jaipur, India, February 1999, pp. 394 - 398.
[13] Petri Bergholm, Mauri Honkanen, Sven-Gustav Häggman, "Simulation of a DS-CDMA Network," *Proc. ICUPC' 95*, November 1995, pp. 838 -842
[14] P.T Brady "A Model for Generating On-Off Speech Patterns in Two-way Conversation," *Bell Syst. Tech. J.*, September 1969, pp. 2445 - 2472.
[15] Mauri Honkanen , Techical Report 31.12.1995. TEKES - Project ,"Simulation and Signal Processing in radio Systms", Sbproject 2.3. "Development of a DS-CDMA Radio Network Simulator." The Ray Tracing Program.
[16] S. Ariyavisitakul, " SIR - Base Power Control in a DS-CDMA System", *Proc. GLOBECOM'95*, November 1995, pp. 838 - 842.

Methods for Measuring and Optimizing Capacity
in CDMA Networks Using Smart Antennas

Scot D. Gordon, Martin J. Feuerstein, Michael A. Zhao
Metawave Communications
PO Box 97069
Redmond, WA 98073
425-702-5884, scotg@metawave.com

Abstract

Smart antennas for CDMA networks have now entered commercial service in a number of IS-95 cellular markets, where they are used to increase capacity through traffic load balancing, managing handoff activity and reducing interference. With the introduction of CDMA smart antennas comes the complex question of how best to first measure and then to optimize capacity. This paper outlines an approach for estimating the forward link capacity of a CDMA cell site using readily available CDMA metrics obtained from switch statistics and drive test data. This capacity model is applied to a real-world smart antenna in a commercial deployment, where a 27% capacity improvement is projected over the existing conventional antenna system

1. Introduction

Due to the explosive growth in the number of digital cellular subscribers, service providers are becoming increasingly concerned with the limited capacities of their existing networks. This concern has led to the deployment of smart antenna systems throughout major metropolitan cellular markets. These smart antenna systems have typically employed multibeam technologies, which have been shown, through extensive analysis, simulation, and experimentation, to provide substantial performance improvements in FDMA, TDMA, and CDMA networks [1-5]. Multibeam architectures for FDMA and TDMA systems provide the straight-forward ability of the smart antenna to be implemented as a non-invasive add-on or appliqué to an existing cell site, without major modifications or special interfaces.

When the question of CDMA smart antennas arises, it is clear from the literature that multibeam techniques lead to significant capacity improvements when the antenna processing is tightly interfaced with, or embedded within, the cell site's baseband receiver processing [4,5]. However, such an architecture lacks the advantage of a simple non-invasive add-on as it requires re-architecting the existing base station infrastructure. More recently, an alternative smart antenna architecture has been proposed for CDMA networks, as is outlined in [8]. This approach synthesizes sector patterns via a phased array antenna providing the capability of creating sectors of varying azimuth, beamwidth and sculpting contours. This flexibility allows one to both evenly distribute the traffic load amongst the sectors and better manage handoffs regions. Further, fine control of the radiation pattern helps mitigate areas of interference and pilot pollution. In effect, the smart antenna approach allows one to maximize the sectorization efficiency.

There have been a number of deployments using this CDMA smart antenna technology, however, quantifying the capacity gains associated with such deployments is a challenging problem. This is primarily because there exists no universally accepted method for measuring the capacity of a CDMA cell site, let alone the capacity of a cell employing smart antennas. Even where theoretical capacity models exist, they are often not parameterized by quantities that can be readily measured in a live, commercial network environment. This paper attempts to bridge this gap by outlining an approach for estimating the capacity of a CDMA cell using simple cell metrics obtained from switch statistics and drive test data. The focus is on the forward link as observations typically show this to be the limiting link in CDMA networks using the 13 kbps vocoder (IS-95 Rate Set 2). Further as an example, results from a sample deployment of a CDMA smart antenna are analyzed and capacity improvements quantified using the forward link capacity estimate.

2. Measuring Forward Link CDMA Capacity

The forward link capacity of a CDMA cell site remains to this day a hotly debated question, with about as many different rules of thumb as there are wireless service providers. From a theoretical perspective the IS-95 forward link capacity is quite difficult to analytically model for several reasons, including imperfections in forward power control, large variations in required C/I for a given voice quality, and strong sensitivity to user locations, handoff states and velocities. Further, these capacity estimation rules of thumb are generally applied network-wide across all cells while the actual forward link capacity varies greatly from sector to sector. We outline an approach to estimate the capacity using simple cell metrics obtained via switch reports or from a mobile diagnostic monitor. We begin by outlining an approach for estimating the forward link capacity of a single sector.

The forward link capacity of a CDMA sector is limited by the need to retain a minimum pilot signal to noise plus interference ratio (energy per chip to noise plus interference spectral density, Ec/Io) for acceptable service. Typically, design guidelines for reasonable downlink performance suggest that the pilot remain at least 15% of total transmit power, so that the maximum transmit power is limited to, Pmax = Ppilot/0.15.

For a CDMA sector the following relationship must hold

$$\phi_p + \phi_{ps} + \sum_{i=1}^{N} \mathbf{v}_i \mathbf{f}_i \leq 1. \qquad (1)$$

That is the fraction of maximum power, Pmax, devoted to the pilot, ϕ_p, plus the fraction of maximum power devoted to the paging and synch channels, ϕ_{ps}, plus the fraction of total available power devoted to each traffic channel at full rate, $\mathbf{f}_i$, times the actual voice activity rate $\mathbf{v}_i$ cannot exceed unity. Here, ϕ_p and ϕ_{ps} are fixed quantities while $\mathbf{f}_i$, $\mathbf{v}_i$ and hence N are random variables. $\mathbf{v}_i$ is discrete assuming values 1, ½, ¼, and 1/8. $\mathbf{f}_i$ is also discrete as there are a finite number of transmit power levels for a traffic channel. It is assumed that $\mathbf{f}_1, \mathbf{f}_2 \ldots$ are independent and identically distributed (iid). The distribution can be determined empirically by tracing a call at the mobile switching center (MSC) and logging the traffic channel power requirements. Figure 1 depicts the fractional traffic power probability distribution for the three sectors at our sample deployment site. The distributions are similar in shape with peaks at the minimum fractional power level allowed by the base station infrastructure equipment.

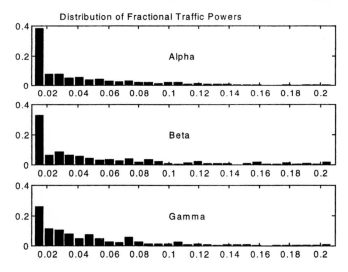

Figure 1: Empirical Probability Distributions for the Fractional Traffic Power.

In practice, certain base station infrastructure implementations block new call originations before the sector reaches its maximum transmit power. The base station does this by monitoring the power requirements of the pilot, paging, synch and all traffic channels and blocks when the total transmit power exceeds a threshold. The threshold is typically selected such that blocking occurs when the sector is transmitting at 85% of maximum power, allowing sufficient headroom for forward power control, variable rate voice and hand-ins.

In order to find the capacity of a given sector we must find the distribution on the number of channels provided by a given sector. The probability that a sector supports at least N channels is

$$\Pr(at\ least\ N\ Channels) = \Pr\left\{ \sum_{i=1}^{N-1} \mathbf{v}_i \mathbf{f}_i < T - \phi_p - \phi_{ps} \right\}, \tag{2}$$

where for an 85% threshold T would be 0.85. The N-1 term is in the limit of the summation because the base station does not block until it exceeds the threshold. Hence, anytime we are under the blocking threshold there exists at least one additional traffic channel available for assignment.

Solving (2) is difficult in that there is no analytic expression for the distribution of the fractional traffic power. An approximate solution is to rely on the central limit theorem and replace the summation in (2) with a Gaussian random variable. If we define

$$\mathbf{X} = \sum_{i=1}^{N-1} \mathbf{v}_i \mathbf{f}_i, \tag{3}$$

the probability distribution of $\mathbf{X}$ is approximately $N\left(m_x, \sigma_x^2\right)$ where

$$m_x = (N-1)m_f\left(pr(\mathbf{v}=1) + \frac{1}{2}pr(\mathbf{v}=1/2) + \frac{1}{4}pr(\mathbf{v}=1/4) + \frac{1}{8}pr(\mathbf{v}=1/8) \right)$$

$$\sigma_x = \sqrt{(N-1)\left\{ \left(\sigma_f^2 + m_f^2\right)\left(pr(\mathbf{v}=1) + \frac{1}{4}pr(\mathbf{v}=1/2) + \frac{1}{16}pr(\mathbf{v}=1/4) + \frac{1}{64}pr(\mathbf{v}=1/8) \right) - m_x^2 \right\}}$$

Here m_f is the mean and σ_f^2 is the variance of $\mathbf{f}_i$. The two quantities can be estimated empirically from drive data. Typically, the four rates can be assumed equally probable so that $pr(\mathbf{v}=1) = pr(\mathbf{v}=1/2) = pr(\mathbf{v}=1/4) = pr(\mathbf{v}=1/8) = 1/4$. With our Gaussian approximation we have

$$\Pr(at\ least\ N\ Channels) \approx \Pr\left\{ \mathbf{X} < T - \phi_p - \phi_{ps} \right\}.$$

Defining

$$Q(x) = \int_x^\infty \frac{1}{\sqrt{2\pi}} e^{\lambda^2/2} d\lambda$$

gives

$$\Pr(at\ least\ N\ Channels) \approx 1 - Q\left(\frac{T - \phi_p - \phi_{ps} - m_x}{\sigma_x}\right). \qquad (4)$$

The probability distribution of N is obtained from

$$P(N) = \Pr(at\ least\ N\ channels) - \Pr(at\ least\ N+1\ channels)$$

Plots of $P(N)$ for our sample deployment site are depicted in figure 2 where we have replaced $P(N)$ with either $P_\alpha(N)$, $P_\beta(N)$, and $P_\gamma(N)$ to denote with which sector these distributions correspond. In the regions where we would expect the Gaussian approximation to be inaccurate (small N) the probability of occurrence is negligible and inaccuracy of the approximation tolerable. Where there exists a significant probability of occurrence, N is large enough that the Gaussian approximation is accurate assuming the independence assumption holds.

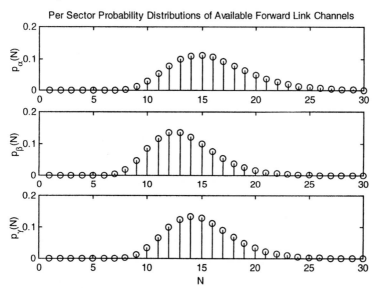

Figure 2: Probability distributions of the number of available forward link channels.

A grade of service (blocking probability) can now be obtained by averaging over the Erlang B formula for each value N. For example, for the alpha sector we have

$$\Pr(Blocking\ in\ \alpha) = P_{B_\alpha}(h_\alpha C) = \sum_{i=1}^{\infty} p_\alpha(i) \frac{(h_\alpha C)^i / i!}{\sum_{j=0}^{i}(h_\alpha C)^j / j!}, \qquad (5)$$

where C is the offered traffic in Erlangs and h_α is the handoff overhead. Inversion of this formula for C yields the sector capacity for a given grade of service. It is worth noting that we are not restricted to the Erlang B formula and could average over any grade of service equation used for a fixed number of channels.

Extension of the above per sector capacity estimate to a cell capacity requires inclusion of traffic distribution information. If traffic is uniformly distributed across all sectors then the resultant cell capacity would be a simple matter of adding the capacity of each of the individual sectors. However, studies of network statistics have shown that the traffic is rarely distributed evenly between sectors. In the worst case scenario all of the traffic load is concentrated on a single sector, yielding a cell capacity that is equal to the capacity of the loaded sector. Incorporating load balance of the cell requires that we include the probability that a user is in a given sector. In doing so our grade of service equation becomes

$$P_B(C) = p_\alpha P_{B_\alpha}\left(p_\alpha h_\alpha C\right) + p_\beta P_{B_\beta}\left(p_\beta h_\beta C\right) + p_\gamma P_{B_\gamma}\left(p_\gamma h_\gamma C\right), \tag{6}$$

where p_α, p_β, and p_γ represent the probability that a mobile is being served by sectors alpha, beta and gamma respectively, h_α, h_β, and h_γ are the softer handoff overheads for sectors alpha, beta, and gamma respectively, and C is the offered traffic to the entire cell. Implicit in (6) is the assumption that the numbers of channels in each sector are independent or, in other words, the joint distribution of available channels for the cell can be factored in to the marginal distribution of available channels for each sector, $p_{\alpha,\beta,\gamma}(N_\alpha, N_\beta, N_\gamma) = p_\alpha(N_\alpha)p_\beta(N_\beta)p_\gamma(N_\gamma)$.

Figure 3 shows the blocking probability versus the traffic intensity as predicted in (6) and as measured from a single day of hourly statistics accumulated at the mobile switching center. We see close agreement from the two with the prediction acting as a very good data trending curve for the observed switch data. Such good agreement between the model and real-world observation provides confidence in the outlined approach for estimating capacity.

As demonstrated by the capacity model, the primary factors influencing forward link cell capacity include the handoff overhead, the load balance of the cell, and the power statistics of the traffic channel. Smart Antennas provide the flexibility to optimize these parameters. Through sector azimuth and beamwidths adjustments the traffic load can be evenly distributed equalizing the utilization of each sector. This flexibility in moving the sector boundaries also allows better management of handoff overhead.

Blocking - Load Balanced Configuration; Measured and Predicted

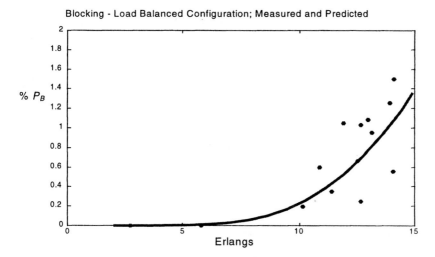

Figure 3: Projected (solid lines) and observed (x's) trend of traffic intensity versus blocking probability.

3. Smart Antenna Deployment Results

A smart antenna system was deployed into a sample cell in a major metropolitan network. We outline the results of this deployment by examining the performance of the specific cell both before and after the presence of the smart antenna. These two configurations are henceforth referred to as the baseline and load balanced configurations respectively. The baseline systems consisted of the existing cell site sector antennas while the load balanced configuration used the smart antenna's phased array antennas to rotate each sector's azimuth pointing angle by 60 degrees. It was determined that this configuration would provide the greatest level of load balance across the sectors for the particular cell site in question, due to the nature of the existing traffic imbalances across the sectors. Table 1 displays the load in terms of the probability that a call is being served by a given sector. An important observation is that the sector with the greatest load in the baseline scenario, beta, was reduced from a probability of 0.49 to 0.38; a 22% reduction. This reduction represents traffic that was offloaded to the adjacent sectors allowing for better sectorization efficiency.

Table 1: Load balance of the baseline and load balanced configurations.

Scenarios	p_α	p_β	p_γ
Baseline	0.22	0.49	0.29
Load Balanced	0.28	0.38	0.34

Table 2 shows the remaining parameters necessary to estimate the blocking probability for the two configurations. This includes the handoff overhead, h, the mean and variance of the traffic channel power, m_f and σ_f respectively, the blocking threshold T, the fractional pilot, paging and synch powers $\phi_p + \phi_{ps}$, and the voice activity. Each of these parameters is grouped under a measurement category to denote how the parameter was obtained.

For those parameters that are not fixed across both configurations, handoff overhead and the statistics of the traffic channel power, the load balancing configuration is as good or better with the exception of the handoff overhead of the beta sector. This is not surprising as the 60 degree rotation had the effect of both load balancing and the movement of handoff boundaries into areas of lesser traffic densities. Further, the handoff boundaries of a phased array antenna are narrower than that of traditional sector antenna because the rolloff of the radiation pattern's main lobe is much steeper with the smart antenna array aperture. If all other things remain the same, this steep rolloff property will nearly always provide some degree of reduced handoff overhead for the smart antenna implementation compared to conventional antennas.

Using the data from tables 1 and 2, figure 4 displays the grade of service versus offered traffic in Erlangs for both the baseline and load balanced configurations. Examining the curves at the 2% grade of service point we observe that the measured capacity of the baseline system is approximately 12.6 Erlangs while the measured capacity of the load balanced configuration is 16 Erlangs; a 27% capacity improvement. Once again, the capacity improvement comes from a combination of traffic load balancing, handoff overhead reduction and per traffic channel transmit power reductions with the smart antenna.

Table 2: Statistics required in blocking calculation for baseline and load balanced configuration.

		Mobile Diagnostic Monitor	Call Trace at the Switch		Cell Site Parameters		Network Statistic
Scenario	Sector	h	M_f	σ_f	T	$\phi_p + \phi_{ps}$	Voice Activity
Load Balanced	Alpha	1.4	0.060	0.044	0.85	0.22	Equally Likely
	Beta	1.8	0.071	0.052	0.85	0.22	Equally Likely
	Gamma	1.8	0.065	0.048	0.85	0.22	Equally Likely
Baseline	Alpha	1.7	0.060	0.046	0.85	0.22	Equally Likely
	Beta	1.6	0.073	0.055	0.85	0.22	Equally Likely
	Gamma	2.2	0.087	0.058 ·	0.85	0.22	Equally Likely

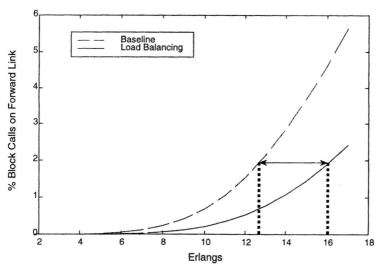

Figure 4: Grade of service for the baseline and load balancing configurations. At a 2% blocking rate there is a 27% capacity improvement offered by the load balancing configuration.

4. Conclusion

A method for measuring the CDMA forward link capacity has been presented and demonstrated to show close agreement with real-world observations. The approach utilizes metrics that are easily obtained through both switch statistics and drive data. The true utility of a capacity estimate is the guidance and insight that it provides in the iterative optimization process to improve capacity. This is particularly true with the advent of smart antennas, which provide arbitrary sector beamwidth, sector azimuths, and sculpted coverage patterns. In the sample deployment scenario described in this paper, the smart antenna was demonstrated to improve the capacity of a cell by 27% at a grade of service of 2%. The capacity increase was achieved by reducing handoff overhead, distributing the load more evenly amongst the sectors, and reducing the power requirements of individual traffic channels.

References

[1] S. C. Swales, M. A. Beech, D. J. Edwards and J. P. McGeehan, "The Performance Enhancement of Multibeam Adaptive Base Station Antennas for Cellular Land Mobile Radio Systems", *IEEE Trans. Veh. Tech.*, Vol 39(1), Feb. 1990, pp. 56-67.

[2] Y. Li, M. J. Feuerstein, D. O. Reudink, "Performance Evaluation of a Cellular Base Station Multibeam Antenna", *IEEE Trans. Veh. Tech.*, Vol. 46(1), Feb. 1997, pp. 1-9.

[3] M. J. Ho, G. L. Stuber and M. D. Austin, "Performance of Switched-Beam Smart Antennas for Cellular Radio Systems", *IEEE Trans. Veh. Tech.*, Vol. 47(1), Feb. 1997, pp. 10-19.

[4] J. C. Liberti, "Analysis of CDMA Cellular Radio Systems Employing Adaptive Antennas", Ph.D. Dissertation, Virginia Tech, Sept. 1995.

[5] J. H. Winters, "Smart Antennas for Wireless Systems", *IEEE Personal Communications*, Vol. 5(1), Feb. 1998.

[6] T. W. Wong and V. K. Prabhu, "Optimum Sectorization for CDMA 1900 Base Stations", *Proc. IEEE VTC'97*, May 4-7, 1997, Phoenix, AZ, pp. 1177-1181.

[7] J. S. Wu, J. K. Chung and C. C. Wen, "Hot-Spot Traffic Relief with a Tilted Antenna in CDMA Cellular Networks", *IEEE Trans. Veh. Tech.*, Vol. 47(1), Feb. 1998, pp. 1-9.

[8] M. J. Feuerstein, J. T. Elson, M. A. Zhao, S. D. Gordon, "CDMA Smart Antenna Performance", *1998 Virginia Tech Symposium*, June 1998, Blacksburg, VA.

Adaptive Radio Resource Control via Cascaded Neural Networks for Sequenced Propagation Estimation and Multi-user Detection in Third-generation Wireless Networks

William S. Hortos

Florida Institute of Technology, Orlando Graduate Center, 3165 McCrory Place, Suite 161
Orlando, FL 32803
Email: hortosw@ccgate.dl.nec.com

ABSTRACT

A hybrid neural network approach is presented to predict radio propagation characteristics and multi-user interference and to evaluate their combined impact on throughput, latency and information loss in third-generation (3G) wireless networks. The three performance parameters influence the quality of service (QoS) for multimedia services for 3G networks. These networks are based on hierarchical cell structures and operate in mobile urban and indoor environments with service demands emanating from diverse traffic sources. Candidate radio interfaces for these networks employ a form of wideband CDMA.

The proposed neural network (NN) architecture allocates network resources to optimize QoS metrics. Parameters of the radio propagation channel are estimated, followed by control of an adaptive antenna array at the base station to minimize interference, and then joint multiuser detection is performed at the base station receiver. These adaptive processing stages are implemented as a sequence of NN techniques that provide their estimates as inputs to a final-stage Kohonen self-organizing feature map (SOFM). The SOFM optimizes the allocation of available network resources to satisfy QoS requirements for variable-rate voice, data and video services. As the first stage of the sequence, a modified feed-forward multilayer perceptron NN is trained on the pilot signals of the mobile subscribers to estimate the parameters of shadowing, multipath fading and delays on the uplinks. A recurrent NN (RNN) forms the second stage to control base stations' adaptive antenna arrays to minimize intra-cell interference. The third stage is based on a Hopfield NN (HNN), modified to detect multiple users on the uplink radio channels to mitigate multiaccess interference, control carrier-sense multiple-access (CSMA) protocols, and refine handoff procedures. In the final stage, the SOFM, operating in a hybrid continuous and discrete space, adaptively allocates resources of antenna-based cell sectorization, activity monitoring, variable-rate coding, power control, handoff and caller admission to meet the QoS for various multimedia services.

The performance of the NN cascade is evaluated through simulation of a candidate 3G network using W-CDMA parameters in a small-cell environment. The simulated network consists of a representative number of cells. QoS requirements for different classes of multimedia services are considered. Initial results show the cascade yields relatively low probability of new call blocking and handoff dropping.

1. INTRODUCTION

The International Telecommunications Union (ITU) has developed requirements, called IMT-2000, for the next generation of mobile communication networks to provide anywhere, any-time, bandwidth-on-demand multimedia services to users. These services include toll-quality voice, variable-rate video, and high-speed data of 144 and 384 kilobits per second (kbps) for high-mobility users with wide-area coverage and 2 megabits per second (Mbps) for low-mobility users with small-cell coverage. As current cellular and PCS digital networks are considered the second generation, IMT-2000 requirements have been created for third-generation (3G) wireless networks. The radio interface design of many IMT-2000 proposals is based on wideband, direct-sequence (DS), code division multiple access (CDMA). A leading proposal, called cdma2000, has been submitted by the CDMA Development Group (CDG) and the Telecommunications Industry Association (TIA) in North America. Another proposal, W-CDMA, is promoted jointly by ARIB in Japan and the European Telecommunications Standards Institute (ETSI).

The objectives of this paper are the estimation and enhancement of system performance in proposed 3G DS-CDMA wireless networks for integrated multimedia services. The approach to these objectives is radio resource allocation (RRA) to effect interference diversity to reduce variance, thereby the fluctuations and increase channel capacity subject to quality of service (QoS) requirements. IMT-2000 requirements offer opportunities to use neural network (NN) techniques in interference-cancelling receiver design. Competing matched filters are often inefficient and offer suboptimal performance in multiuser detection (MUD). Other near-far resistant receivers are too complex. Since a DS-CDMA system is interference limited, properly designed interference-cancellation methods improve capacity. Thus, adaptive power control and other interference-mitigation techniques based on NN techniques are applied to improve signal-to-interference ratio (SIR). Effects of interference variation on the QoS of integrated services with different rates and powers has recently been considered, but not in great depth.

2. FEATURES OF THIRD-GENERATION DS-CDMA NETWORKS

The 3G air interface proposals based on CDMA focus on two main types, network asynchronous and synchronous. In the former type, the base stations (BSs) are not synchronized, while in the latter they are synchronized within a few microseconds. The W-CDMA system is an asynchronous network. The length of this paper limits the focus to salient features of W-CDMA uplinks.

W-CDMA radio links offer variable bandwidths of 1.25, 5.0 MHz and higher multiples of 10 and 20 MHz in the future systems. Chip rates are 1.024, 3.840 Mcps, and later 2×3.840 Mcps and 4×3.840 Mcps. W-CDMA employs long spreading codes.[1] Variable-length orthogonal sequences are used as channelization codes. A short variable-length Kasami code is proposed on uplinks for MUD implementation. On uplinks W-CDMA employs time-multiplexed pilot symbols for coherent detection. The user-dedicated pilot symbols can be used for channel estimation with adaptive antennas as well.

The W-CDMA traffic channel structure is based on a single-code transmission for low-data rates and multicodes for higher rates. Multiple services belonging to the same connection are time-multiplexed in stages. Time multiplexing occurs after both outer coding and inner coding; the multiservice data stream is mapped to one or more dedicated physical data channels. In multicode transmission, data channels are alternately mapped into the quadrature (Q) channel or the in-phase (I) channel.

W-CDMA has two different types of packet data transmission methods. Short data packets can be appended directly to a random access burst. This method, called *common channel packet transmission*, is used for infrequent packets, where link maintenance for a dedicated channel (DCH) would result in unacceptable overhead. Longer, more frequently transmitted packets are sent on DCHs. A large single packet is transmitted using a single-packet scheme, where the DCH is released immediately after packet transmission. In a multipacket scheme the DCH is maintained by transmitting power control and synchronization information between subsequent packets. The random access burst is 10 ms long and transmitted with fixed power. Access is based on the slotted Aloha scheme. Data arrives on the transport channel as *transport blocks*. A variable number of transport blocks arrive on each transport channel at each transmission time instant. The *transmission time interval* is restricted to the set {10, 20, 40, 80 ms}.

A key W-CDMA feature is the transmission of multiple parallel services (transport channels) with different QoS requirements on one connection. Parallel transport channels are separately channel-coded and interleaved. Coded transport channels are then time-multiplexed into a coded composite transport channel. Different coding and interleaving schemes can be applied to a transport channel depending on the specific QoS requirements for error rates, delay, etc. Rate matching is applied to reconcile the bit rate of the coded composite transport channel to one of the limited set of bit rates assigned to the physical channels. Static rate matching is distributed between parallel transport channels so that transport channels fulfill their QoS requirements at approximately the same channel signal-to-interference ratio (SIR).

3. QUALITY OF SERVICE IN WIRELESS MULTIMEDIA SERVICES

3G multimedia services can be classified into two main categories: real-time and packet data. Real-time services can be variable-rate, e.g., the 8-kbps and 13-kbps voice codecs used in IS-95. In real-time mode, a large amount of digitized information is transmitted over a relatively long duration, whereas packet-data services are provided to bursty information sources characterized as on-off processes. For packet-data services, transmission stops at the end of the data burst, with no information generated during the unpredictable off intervals. Real-time services may be selected as constant bit-rate (CBR) or variable bit-rate (VBR), and transmission is continuously maintained during the call. Packet-data services are provided to users with demand for high transmission rates, but short service times.

3.1. W-CDMA Operation for Multimedia Services

W-CDMA offers a number of options to integrate multirate services: (1) trade off processing gain for an increased information rate in the same spread bandwidth and (2) pair up basic data channels until the required information rate is obtained. The phrase "basic channel" refers to CBR transmission with the highest processing gain. The radio resource controller fully directs the choice of appropriate coding scheme, interleaving, and rate-matching parameters.

The media access controller (MAC) must support a mixture of services. The MAC protocol controls the data stream delivered to the physical layer over the transport channels. If an MS wants to transmit data of different services, e.g., a real-time service and packet data, it is assigned two sets of transport formats, one for real-time and one for packet data. As for a single service, the MS may use any transport format assigned for real-time services, but may only use transport formats specific to packet data. The MS is assigned a specific output power/rate threshold. The aggregate output power/rate can never exceed the threshold. Thus, the transport formats used for data service fluctuate adaptively to the transport formats used for speech service.[2] One proposed handoff approach dynamically adapts the amount of applied RRs based on current network conditions, that is, on the average connection-dropping probability and utilization of RR reserves, to improve the RR utilization as well as the blocking probability.[3]

3.2. Models of Interference, Multimedia Service Demand, and Radio Resource Allocation

The following conditions express the interference and QoS constraints in the operation of wireless multimedia networks. The model's dimensions are larger than the exposition in [4] due to an increase in the number of RR categories available in W-CDMA to support demands for simultaneous multiple services.

1. Co-channel constraint (CCC). The same transport (physical-layer) channel cannot be assigned simultaneously to certain pairs of mobile users in the cells. The CCC is determined by co-channel interference (CCI), which depends on the interference control applied at the N base stations (BSs) of the network.

2. Adjacent channel constraint (ACC). Channels adjacent in their domain's distance metric (frequency, time slot or PN code) cannot be assigned to adjacent radio cells simultaneously.

3. Co-site channel constraint (CSC). Any pair of channels assigned to a radio cell must be at a minimum distance in their domain. In W-CDMA, minimum distance depends on the interference level produced by adaptive antenna selection, activity monitoring, power control and service classes active in each BS's coverage area.

The constraints have previously been described for single-service networks by an $N \times N$ symmetric matrix, called the *interference matrix* C. Each off-diagonal element c_{ij} in C represents the minimum separation distance between a channel assigned to cell (or sector) i and a channel assigned to cell (or sector) j. The CCC is represented by $c_{ij} = 1$, while the ACC is represented by $c_{ij} = 2$. Setting $c_{ij} = 0$ indicates that BSs i and j are allowed to assign the same channel to users in their service areas. Each diagonal element c_{ii} in C represents the minimum separation distance between any two channels assigned to cell (or sector) i. This is the CSC and $c_{ii} \geq 1$ is always satisfied, provided that, in sectored networks, sectors are equivalent to cells. In 3G networks, the matrix dimensions increase to accommodate the number of physical-layer channels available to each BS to support multiple real-time and packet-data services to each active MS. Let C be the maximum number of physical-layer channels that can be supported at any BS. Then, the 3G W-CDMA interference matrix is an $N \cdot C \times N \cdot C$ symmetric matrix.

DS-CDMA capacity can only be increased by reducing other-user interference I. This fact suggests a departure from the model of a two-dimensional interference matrix for N BSs assigning fixed M channels. In W-CDMA each BS i is assumed to be able support $m_i^c(I)$ common channels and $m_i^d(I)$ dedicated channels, where $m_i^c(I) + m_i^d(I) = m_i(I)$ and $m_i(I) \leq M(I)$, the total number of channels in the network, and $m_i(I)$ represents the local channel capacity of the service area i and $M(I) \leq N \cdot C$. As discussed by Gilhousen, *et. al.*[5], using adaptive antenna-array beamforming, voice activity monitoring, selectable spreading factors and coding rates, and power control can regulate interference, I, in a W-CDMA network. Interference regulation determines the number of available channels. Viewing the adaptive elements as the RR controls of the network, their effect on channel capacity is represented as a composite mapping, $\Psi = \Phi \bullet \Gamma$, on a $4N$-dimensional lattice, $\Re : \Phi : \Re = (S \times V \times B \times P)^N \rightarrow I^N; \Gamma : I^N \rightarrow M = \{m : m = (m_1, m_2, \ldots, m_N); m_i = \#$ channels in the service area of BS $i\}$, where S represents the set of antenna-array sector values in any BS coverage area; V is the set of states of voice activity monitoring in each area; B is the set of spreading factors $(4 - 256)$ and/or coding rates $(1, 2, 4, 8, 13, 32$ kbps); P is the set of discrete power control levels to the MSs $(0 - 10$ dB, in 0.25 dB steps); I is the real interval bounding interference levels, while M is the subset of the set of N-dimensional vectors of non-negative integers, whose i'th component is the total channel capacity at BS i. The composite mapping relates the RRA to channel capacity, through the interference level that the assignment generates in network cells.

For a single-service network, the traffic demand for physical-layer channels in each BS coverage area, in a network of N BSs, is represented by an N-vector called the *traffic demand vector* T.[4] Simultaneous multimedia services increase traffic demand dimensions to form an array T, where each row vector $t_i = (t_{i1}^{rt}, t_{i2}^{rt}, \cdots, t_{iK_i}^{rt}; t_{i1}^{p}, t_{i2}^{p}, \cdots, t_{iL_i}^{p})$, with $t_{ij}^{rt}(t_{ij}^{p})$ the number of units of real-time (packet-data) service class j from both new calls and handoffs assigned to cell i. The t_{ij}^{rt} and t_{ij}^{p} are nonnegative with a zero value indicating no service demands of class j at cell i. Let q_{ikl} denote the physical-layer channels to support service class l units of the k'th active call assigned to cell i. Then, the interfering channel constraints described by the interference matrix can be expressed by the relations:

$$|q_{ikl} - q_{jmn}| \geq c_{ij}, \text{ for } i \neq j, \ 1 \leq i, j \leq N \cdot C; \ k \neq m, \ 1 \leq k \leq t''_{ii}, \ 1 \leq m \leq t''_{jn}, \text{ where } u = rt \text{ or } p. \quad (1)$$

Since the same physical-layer channel cannot be used simultaneously in two interfering cells (intercell) or by two interfering users (intracell), interference conditions have been considered hard constraints. With the introduction of adaptive RRs, the constraints can be considered "soft" to the limits imposed by bounds on the resource sets. When service demands cannot be satisfied due to a constraint, the corresponding request for a new call or a call handoff is blocked. For this reason, the joint probability of call blocking or handoff dropping is a useful performance metric for an adaptive RRA algorithm.

4. THE CASCADE OF NEURAL NETWORKS

NN estimation techniques can be applied to multipath fading, imperfect power control, and non-uniform traffic. Multipath fading is a major impairment to CDMA operation, since each additional path adds extra interference.[6] To support integrated multimedia services, multiaccess interference (MAI) at the BS requires mitigation to meet QoS requirements.[7] MAI reduction greatly increases link capacity. A candidate method is the interference canceller; another is the adaptive antenna array, viewed as adaptive cell sectorization. A third is MUD, based upon NN techniques, such as reduced-complexity radial basis functions (RBFs) or Hopfield NNs (HNNs). Interference cancellers can be essentially classified as either single-user or multiuser. The former reduces MAI using a linear filter in one instance, and is simpler to implement than the latter. W-CDMA uses long PN spreading code sequences on the uplink. The time-varying nature of such a code sequence, when observed over every symbol period, excludes adoption of single-user interference cancellers. The decorrelating MUD receiver requires very complex computation of the inverse correlation matrices among different users' spreading codes and is considered impractical. The multi-stage IC version of the non-linear replica generation type is attractive, since interference replica generation and subtraction is performed successively for different users.[8] MUD receivers require knowledge on users' parameters, e.g., time delays, signal strengths, etc. Accurate channel estimation is needed to generate the interference replica of each user.

Feedforward multilayer perceptron (MLP) neural networks (NNs) are proposed to estimate signal strength, fade rate, shadowing standard deviation, and principal path coefficients on uplinks, based on the received pilot signals from mobiles. Outputs from the MLP-NN in addition to service demands of the active calls and handoffs are input to a second-stage recurrrent neural network (RNN) for adaptive antenna array control. The RNN produces estimates of the number and type of beams or sectors, denoted $\hat{s}_i$, and the directed gain of each active element at BS i, $\hat{g}_{ik}$, $k = 1, \ldots, \hat{s}_i$. These estimates along with service demands of calls and handoffs are input to the third-stage MUD receiver, based on a discrete-form HNN to reduce MAI. The reduced interference eases requirements for stringent power control, antenna control, activity monitoring, and code and spreading-factor assignments to support the QoS for multimedia service demands. It also increases the reserve of available RRs, thereby reducing the possibility of new call blocking; unnecessary handoffs, both soft and hard; and dropped handoffs to other BSs. Excessive call handoffs are a major component of end-to-end connection latency that impacts QoS in real-time services. The reduced MAI, antenna-array sector indices and gains, generated in the first three NN stages, along with service demands, are collectively input to a modified Kohonen SOFM. The SOFM determines the best RRA array, mapping through the estimated residual MAI profile to meet the current multimedia service demands. This is shown in Figure 1.

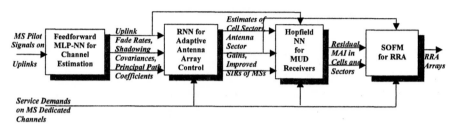

Figure 1. Cascade of Neural Network Stages for Adaptive RRA

4.1. Multilayer Perceptron Neural Network for Propagation Channel Estimation

The first stage of the cascade is a fast NN method for uplink propagation channel estimation in an urban environment. It extends the method in [9]. The estimation error and computation time of the technique is compared to that of COST 231 models [10], Walfisch-Bertoni model [11], and the Saunders-Bonar model.[12]

Propagation losses between two points can be expressed as a sum of free-space path losses L_0, which depend on frequency f and distance d and an attenuation term for the effect of shadowing. The factor ρ is the fade rate determined by multipath fading and σ is the excess attenuation term due to shadowing:

$$L = L_0 + \sigma_{obs} = 32.4 + 20\log[f(\text{MHz})] + 2\rho\log[d(\text{km})] + \sigma_{obs} \qquad (2)$$

The attenuation term is a function of the heights and spatial distribution of the buildings and other man-made obstructions between the MS transmitter and BS receivers. Electromagnetic models provide accurate predictions, but suffer from long computation.[11, 12] The uplink from MS j to BS i is modeled by the transfer function, $\lambda_{ji}(z) = \sum_{l=0}^{U} \lambda_{jil} z^{-l}$, where U, the number of principal paths, is typically less than 8.

The MLP-NN has been trained and tested with actual BS site measurements taken in Munich as well as simulated pilot signals varied over W-CDMA design parameter values. MLP-NN predictions reported in [9] show a mean error in the test sets of –2.1dB and 0.44 dB and a standard deviation of error of 6.3 dB and 6.6 dB, respectively. These results have at least 50% lower standard deviation compared to predictions given in the COST 231 final report.[13] Continuous operation of the BSs allows frequent updates of the MLP-NN estimates to improve accuracy and to adapt to changing conditions.

The MLP-NN approach reduces the computation time of channel predictions in network planning. Propagation losses have been predicted for a 3392×2400 m^2 urban area with a resolution of 4×4 m^2 using the three models cited above and the MLP-NN.[9] Results indicate for this area size that the MLP-NN is at least four times faster than any cited method. Further speed improvements can be achieved by implementing the NN in parallel analog hardware and by scaling its operation over smaller 90×90 m^2 picocells with a resolution of 1×1 m^2 to achieve at least 62 times reduction.

4.2. Recurrent Neural Network for Adaptive Antenna Control

The second NN stage implements space- and time-diversity combining of mobile signals on uplinks via adaptive antenna arrays using a recurrent NN (RNN) method. The RNN offers better performance with lower complexity than adaptive arrays based on square-root recursive least squares (RLS) schemes.[14]

The RNN method is evaluated for a QPSK/W-CDMA system with L receiving antenna elements, over time-varying multipath channels. Following QPSK demodulation in the BS receiver, the pilot channels are despread with matched filters to obtain complex signals $\hat{x}_1(n),...,\hat{x}_L(n)$. Since the RNN accepts only real inputs, complex signals are formatted into I and Q components for RNN input. Thus, for L receiving antennas, the RNN has $L_0 = 2L$ external inputs and $L_1 = 2$ fully interconnected neurons, providing I and Q output signals. The neuron output at time $n+1$ depends on the external inputs $u_l(n)$ at the previous time instant and the previous outputs of the neurons $v_l(n)$, described by the following:

$$s_k(n+1) = \sum_{l=1}^{L_1} w_{kl}(n)v_l(n) + \sum_{l=1}^{L_0} w_{k,l+N_1}(n)u_l(n) \tag{3}$$

$$v_k(n+1) = f(s_k(n+1)) \tag{4}$$

where $w_{kl}(n)$ is the weight of the connection from l'th input to k'th neuron and $f(\cdot)$ is the sigmoid function.

A widely known algorithm for training RNNs is the real-time recurrent learning (RTRL) algorithm, which updates RNN weights according to the following rule.[15] For $i, k = 1, ..., L_1$ and $j = 1, ..., L_0+L_1$,

$$w_{ij}(n+1) = w_{ij}(n) + \alpha \sum_{k=1}^{L_1} e_k(n+1)p_{ij}^k(n+1) \tag{5}$$

$$p_{ij}^k(n+1) = f'(s_k(n+1))\left[\sum_{l=1}^{L_1} w_{kl}p_{ij}^l(n) + \delta_{ik}u_j(n) \right] \tag{6}$$

where α is the learning gain, $e_k(n) = d_k(n) - v_k(n)$ is the error at the k'th neuron, d_k is the desired output, δ_k is the Kronecker delta , and $f'(\cdot)$ the derivative of the sigmoid $f(\cdot)$. Algorithm "sensitivity" is defined as $p_{ij}^k = \Delta v_k / \Delta w_{ij}$. The RNN is trained with random pilot symbols and weights w_{ij} initialized to random values satisfying $\left| w_{ij} \right| < 10^{-2}$.[14] The learning gain α values for the RTRL algorithm are selected between 0.04 and 0.1 after heuristic optimization.[14] Following training, the RNN is set to decision-directed mode to track the channel variations and correct for distortions on the transmitted pilot signals.

The RNN performance has been compared to that of a linear adaptive array structure trained with the RLS algorithm in an IS-95 CDMA network.[14] The linear structure has a two-tap FIR filter with complex coefficients in each antenna branch, and a training period of 200 pilot symbols. With seven co-channel interferers, at a BER of 10^{-3}, the RNN structure with two receiving antennas performs 3 dB better than the RLS technique. With four receiving antennas, improvement increases to 4 dB. A six-element RNN array performs slightly better than the RLS technique, with a maximum improvement of 2 dB at a BER of 10^{-5}. The BER performances of the RNN and RLS technique are compared as the number of mobile users varies.[14] With SNR fixed at 14 dB in an IS-95 system and four receiving antennas, the RNN performs about seven orders of magnitude better than the RLS technique for a single user. As the number of users increases, the relative advantage of the RNN over the RLS decreases to about 2.5 orders of magnitude for 16 users. The result shows that, while RNN arrays have an advantage for channels dominated by multipath fading, only smaller improvements are obtained in interference-dominated channels. This is the rationale for high-level interference mitigation in the third-stage MUD NN of the cascade.

4.3. Hopfield Neural Networks for Joint Multiuser Detection

In DS-CDMA each MS transmits a different signature waveform, known to the BS receiver. The received signal at the BS is the superposition of signals transmitted by each individual MS. As shown by Verdu, in both synchronous and asynchronous transmission cases, optimal multiuser detection (OMD) is an NP-hard problem, equivalent to maximizing an integer quadratic objective function.[16] Mitra and Poor in [17] have proposed receivers based on RBFs, whose output is a linear combination of nonlinear functions, each of which is applied to the vector input data. These RBF receivers are useful for decentralized detection of a single-user in multiuser channels. They perform well for a small number of synchronous users, but training time is exponential in the number of users.

Kechriotis and Manolakos have introduced the design of a single-layer feedback NN receiver with $O(K)$ neurons, capable of demodulating information transmitted by K synchronous or asynchronous users, sending CDMA packets over the same nearly Gaussian channel. Since OMD can be formulated as an energy minimization problem, it can thus be solved in practically constant time using an analog VLSI-implemented HNN.[18, 19] Simulation suggests the HNN detector outperforms conventional matched-filter detectors to attain near-optimal BER performance with lower complexity than RBF detectors.

If coded waveforms assigned to each mobile user are orthogonal and transmitted signals are antipodal ($\{+1,-1\}$), the conventional detector (CD) can recover information bits by first passing the received signal through a bank of filters matched to the users' signature waveforms, then deciding on bits based on the sign of the output. However, CD performs poorly when powers of the transmitting users are dissimilar.

Assume K active transmitters share the same channel at a given time. A signature waveform $s_k(t)$, limited to $t \in [0,T)$, is assigned to each transmitter. Denote the i'th information bit of the k'th user as $b_k^{(i)} \in \{+1,-1\}$. In a DS-CDMA system, the signal at a receiver is the superposition of K transmitted signals and additive noise. Each $s_k(t)$ is the convolution of the transmitted MS traffic channels and the mulitpath-channel transfer function, estimated in the first MLP-NN stage.

$$r(t) = \sum_{i=-P}^{P} \sum_{k=1}^{K} b_k^{(i)} s_k(t - iT - \tau_k) + n(t), \ t \in \mathbf{R} \tag{7}$$

In (7) $\tau_k \in [0,T)$ are the relative time delays between the users and $2P+1$ is the packet or frame size. In systems where the BSs cooperate to maintain synchronism, $\tau_k = 0$, $k = 1,...,K$. In a CD a simple thresholding device produces an estimate $\hat{b}_k^{(i)}$ for the i'th information bit of the k'th user based on the sign of the i'th output of the k'th matched filter:

$$y_k^{(i)} = \int_{iT-\tau_k}^{(i+1)T-\tau_k} r(t)s_k(t - iT - \tau_k)dt \ , \ \bar{b}_{CD}^{(i)} = sign(\bar{y}^{(i)}) \tag{8}$$

where $\bar{y}^{(i)} = [y_0^{(i)}, y_1^{(i)}, \cdots, y_{K-1}^{(i)}]$. An OMD estimate is produced for the information vector transmitted at the discrete instant i, based on the maximization of the logarithm of the likelihood function. In the synchronous case, it holds that:[20]

$$\bar{b}_{OMD}^{(i)} = \arg \max_{\bar{b} \in \{+1,-1\}^K} \{2\bar{y}^{(i)^T} \bar{b} - \bar{b}^T H \bar{b}\} \tag{9}$$

where $H \in \mathbf{R}^{K \times K}$ is the symmetric matrix of signal cross-correlations, $h_{kl} = \int_0^T s_k(t)s_l(t)dt$.

A detection scheme, suboptimal to the OMD, with low computational complexity called the multistage detector (MSD) has been proposed.[21] The MSD consists of a sequence of stages $m = 1, 2, \ldots,$ each producing an estimate $\bar{b}_{MSD}^{(i)}(m)$ given as:

$$\bar{b}_{MSD}^{(i)}(m+1) = sign\left(\bar{y}^{(i)} - (H - E)\bar{b}_{MSD}^{(i)}(m)\right) \tag{10}$$

where E is a diagonal matrix with elements $e_{ii} = \int_0^T s_i^2 dt$ (signal energies). The output of the $m=1$ stage is initialized to the estimate of the CD. The MSD is insensitive to the near-far problem. In the *asynchronous* case, the OMD problem is written in the form of (9), defining matrices $H(j) \in \mathbf{R}^{K \times K}$, $j = -1, 0, 1$ as $h_{kl}(j) = \int_{-\infty}^{\infty} s_k(t - \tau_k) s_l(t + jT - \tau_l) dt$ and the matrix $\tilde{H} \in \mathbf{R}^{(2P+1)K \times (2P+1)K}$ as

$$\tilde{H} = \begin{bmatrix} H(0) & H(-1) & 0 & \cdots & 0 \\ H(1) & H(0) & H(-1) & & \vdots \\ 0 & H(1) & H(0) & \ddots & 0 \\ \vdots & & \ddots & \ddots & H(-1) \\ 0 & \cdots & 0 & H(1) & H(0) \end{bmatrix}. \tag{11}$$

The asynchronous optimum receiver is viewed as a larger combinatorial optimization problem of the form

$$\tilde{b}_{OMD}^{(m)} = \arg \max_{\bar{b} \in \{+1, -1\}^{(2P+1)K}} \{2 \tilde{y}^{(im)^T} \tilde{b} - \tilde{b}^T \tilde{H} \tilde{b}\} \tag{12}$$

where $\tilde{y}^{(m)^T}$ is the row vector consisting of the sampled outputs of the matched-filter bank corresponding to the m'th packet. When packet length is relatively large, even a small number of users cause a restrictive computational effort to solve (12).

HNNs are single-layer networks with output feedback consisting of simple processors (neurons) where the connection between two processors is established through a conductance T_{ij} that transforms the voltage outputs of amplifier j to a current input for amplifier i. Externally supplied bias currents I_i are also input to every neuron j. Each neuron updates its activation according to the rule: [22]

$$V_i = g(U_i) = g\left(\sum_{j \neq i} T_{ij} V_j + I_i\right) \tag{13}$$

where $g(U_i)$ can be an antipodal thresholding function resulting in $V_i = g(U_i) = sign(U_i)$. Hopfield has shown that, for symmetric connections ($T_{ij} = T_{ji}$), the activation equations (13) always lead to convergence to a stable state.[22] Moreover, when the T_{ii} are zero and $g(\cdot)$ approaches the antipodal thresholding function, the stable states of the network of N neurons are local minima of the energy function given as:

$$E = -\frac{1}{2} \sum_{i=1}^{N} \sum_{j=1}^{N} T_{ij} V_i V_j - \sum_{i=1}^{N} V_i I_i \tag{14}$$

The cross-correlation matrix H is symmetric. Moreover, equation (9) can be rewritten as

$$\begin{aligned} \bar{b}_{OMD}^{(i)} &= \arg \min_{\bar{b} \in \{+1, -1\}^K} \{-\bar{y}^{(i)^T} \bar{b} + \frac{1}{2} \bar{b}^T H \bar{b}\} = \arg \min_{\bar{b} \in \{+1, -1\}^K} \{-\bar{y}^{(i)^T} \bar{b} + \frac{1}{2} \bar{b}^T (H - E) \bar{b} + \frac{1}{2} \bar{b}^T E \bar{b}\} \\ &= \arg \min_{\bar{b} \in \{+1, -1\}^K} \{-\bar{y}^{(i)^T} \bar{b} + \frac{1}{2} \bar{b}^T (H - E) \bar{b}\} \end{aligned} \tag{15}$$

since $\bar{b}^T E \bar{b}$ is always a positive number. The OMD objective function can be translated into the HNN energy in (14) with symmetric weight matrix $T = -(H - E)$ with zero diagonal elements and biases $I = \bar{y}^{(i)}$.

The initial state of the HNN MUD coincides with the initial state of the CD. Active users are assumed to vary relatively slowly and can be estimated by the MLP-NN and RNN stages. The HNN weights can be preset according to the users' energies and the known values of the cross-correlations of their signature waveforms. The discrete-time approximation of the equation of motion of the i'th neuron of the HNN is given by $\Delta U_i = -\dfrac{U_i}{\tau} + \sum_{i \neq j} T_{ij} V_j + I_i$. If $g(\cdot) = sign(\cdot)$, the dynamics of the i'th neuron at the instant $t = m + 1$, are described by the following:

$$V_i(m+1) = sign(U_i(m) - \frac{U_i(m)}{\tau} + \sum_{i \neq j} T_{ij} V_i(m) + I_i) \qquad (16)$$

Setting $\tau = 1$ and substituting in equation (16) for the values of T and I for the proposed HNN detector, (16) becomes $V_i(m+1) = sign\left(y_i - \sum_{i \neq j} h_{ij} V_j(m)\right)$. This last term can be written in matrix form:

$$V(m+1) = sign(\bar{y} - (H - E)V(m)) \qquad (17)$$

Computing (17) and (10) for RC constant $\tau = 1$ and $g(\cdot) = sign(\cdot)$, the $(m+1)$'th stage estimate of the MSD coincides with the output of the discrete-time approximation of this HNN at $t = m + 1$. Since the update of each MSD stage is performed synchronously, an infinite number of MSD stages is essentially equivalent to a discrete HNN operating in synchronous, fully parallel updating mode.[22] Under certain conditions, the HNN energy function has a unique local minimum that coincides with the global minimum of the OMD problem.

In the asynchronous case, the dimension of the optimization problem grows dramatically with packet size and the number of users. If K users transmit packets of length $2P + 1$, the corresponding HNN receiver has $K \cdot (2P + 1)$ neurons. Due to sparsity of $\tilde{H}$, the number of interconnections required for the HNN is reduced to $(3(2P + 1) - 2) \cdot K^2$. When packet size is relatively small and K is small to moderate, an extended version of the HNN detector used in the synchronous case can be used. Computer simulations for $K = 3$ asynchronous users transmitting packets of length 31 bits have been performed.[18] Due to the large OMD detector size and long simulation time, comparisons are reported only with respect to the CD and HNN detector with $g(\cdot) = sign(\cdot)$ and $\tau = 1.0$.

BER vs. SNR is compared for $K = 3$ asynchronous users using optimized Gold sequences of length $L = 127$. The energy of one user is 10 times larger than the energy of each of the other two, so that the maximum near-far ratio is 10. Packet length is equal to 31 bits. The cumulative BER has been computed by simulating both the CD and HNN MUD for 10^7 transmitted sets of symbols for each SNR value, randomly drawn from a uniform distribution.[18] Results for this case show the HNN detector to have uniform improvement in SNR over the BER range from 10^{-2} to 10^{-4} of 1 to 1.5 dB compared to CD performance. During simulations, values of the delays τ_k, $k = 1, 2, 3$, are changed randomly every 500 symbols, so that BER values represent performance of the detectors averaged over all possible delays.

4.4. Self-organizing Feature Maps for RRA

The notion of RRs in a W-CDMA network "competing" to be assigned calls suggests application of the SOFM approach for the last stage of the cascade. The approach modifies Kohonen's SOFM to solve discrete-space optimization problems among the lattice of RRs.[23] Development of a static RRA (SRRA) and extensions to dynamic RRA (DRRA) problems are discussed in previous work by the author.[4] All feasible solutions to the SRRA problem lie at the vertices of an n-dimensional hypercube, where $n = N \cdot R$, N is the number of BSs and $R = S \cdot V \cdot B \cdot P$ available RR combinations of antenna beams or sectors, activity monitoring, coding rates/spreading factors, and power control. Note n is the dimension of $\Re$, the domain of Ψ. The Ψ-image of the vertices also intersects the constraint hyperplane defined by the interference matrix, traffic demand array and channel constraints (1) due to the RRAs. Since each entry t_{ij}, of the traffic demand array T can be assumed integer-valued for all i, j, the image of the RR constraints set can be shown to form an integral polytope. Neurons on this hypercube are defined as

$$X_{j, \Psi(r)} = \begin{cases} 1, \text{ if coverage area } j \text{ is assigned physical channels } \Psi(r) = \Psi \ (s, v, \beta, \ p) \\ 0, \text{ otherwise} \end{cases}$$

for $j = 1, \cdots, N$; and $r \in S \times V \times B \times P$. For convenience, they are denoted by $X_{j,r}$. Let X denote the n-dimensional array of these variables. Normalizing the range of values for each RR and the interference bounds to the interval $[0,1]$, the set of RRs and its Γ-image in the interference range are each contained in unit hypercubes. A vertex is approached continuously from within the unit hypercube, starting from a point on the constraint hyperplane and inside the hypercube. This represents a feasible, non-integer solution to the RRA problem. The continuous variable approach in the interior of the hypercube is denoted by $w_{r,j}$, so that, for a quality metric Q, $Q(W) = Q(X)$ at the vertices. The value $w_{r,j}$ represents the *probability* that the variable in the r, j position of the array X is activated. The vector r is integer-valued, an index into the lattice of allowable RRAs. Kohonen's self-organization is applied to the array of *synaptic weights*, W. This modification permits the SOFM to solve discrete-space optimization problems.

The structure of the discrete-space SOFM consists of an input layer of N nodes, and an $R \times N$-dimensional array of output nodes. The output nodes correspond to the solution array of discrete-valued RRAs, while the input layer represents the N BS coverage areas in the W-CDMA network. The weight connecting input node j to node r of the output array of nodes is given by $w_{r,j}$. A cell in which an assignment of r is required is presented to the network through the input layer at node j. Physically, an incoming call or handoff is presented to the network at BS j. Nodes of the output layer compete with each other to determine which subarray of the solution array meets the QoS requirements of the input with minimal impact on the cost potential. The synaptic weights are then *adapted* to indicate the RRA decision using the neighborhood topology.

Consider the case where RRA r is required at BS j^*. An input vector x is presented to the network with a "1" in position j^* and 0 elsewhere. For each node $r = (i_1, i_2, \cdots, i_R)$ of the outer layer, the value V_{r,j^*}, the cost to the objective function of RRA r to BS j^*, is computed. The *cost potential* V_{r,j^*} of node r for a given input vector x ($x_j = 0, \forall j \neq j^*, x_{j^*} = 1$) is defined by

$$V_{r,j^*} \equiv \sum_{i=1}^{N} \sum_{s \in \Re} P_{j^*,i,(\|\Psi(r)-\Psi(s)\|+1)} W_{i,s} \tag{18}$$

where the interference caused by the RRA is represented by the weight $P_{j,i,d+1}$, termed the proximity indicator where $d = \|\Psi(r) - \Psi(s)\|$ is the distance in the service-channel capacity range between the images of RRAs r and s. If $\Psi(r) = \Psi(s)$, $j^* \neq i$, then interference cost should be at a maximum, with cost decreasing until the two active channels are sufficiently separated, so that MAI, or contention for resources, is below a threshold value. The array P is defined as

$$P_{j,i,d+1} = \max(0, P_{j,i,d-1}), d = 1, \cdots, M-1; \quad P_{j,i,1} = c_{ji}, \forall j, i \neq j; \quad P_{j,i,1} = 0, \forall j. \tag{19}$$

The *dominant node*, m_0, of the outer layer is the node with minimum cost potential for a particular input vector. In the terminology of this representation $V_{m_0 j^*} \leq V_{r,j^*}$ for all nodes r and fixed j^*. The *neighborhood* of the dominant node m_0, is the set of nodes $m_1, m_2, \cdots, m_{\eta_{j^*}}$, ordered according to the values of the cost potentials, i.e., $V_{m_0 j^*} \leq V_{m_1 j^*} \leq V_{m_2 j^*} \leq \cdots \leq V_{m_{\eta_{j^*}} j^*}$, where η_{j^*} is the size of the neighborhood in the SOFM network for BS j^*. Thus, dominant nodes and their neighborhoods are determined by competition according to the objective function, and the weights are modified according to Kohonen's weight adaptation rules within the dominant neighborhood.

When weight updating is complete, the array W has been moved in a direction that may be away from the constraint hyperplane, resulting in an infeasible solution. In the next step, the weights of the nodes outside the dominant neighborhood organize themselves around the modified weights, so that W remains a feasible solution to the RRA problem during the update. This step can be performed by a hill-climbing HNN or HC-HNN. Representing the weight matrix W as a vector w, w is considered to the vector of states of a continuous HNN. The HNN performs random and asynchronous updates on w, *excluding* weights in the dominant neighborhood, to minimize the energy function:

$$E \equiv \|w - (\wp w + \tau)\|^2 \tag{20}$$

where $\wp$ is the projection onto the constraint hyperplane given by

$$\sum_{r \in \mathfrak{R}} X_{j,r} = t_j \ , \forall j = 1, \cdots, N. \tag{21}$$

and $\tau = (\mathbf{I} - \wp)T$, where $\mathbf{I}$ is the identity operator. The energy (20) is expressed in terms of a solution vector x, constructed from the solution array X, by ordering the elements $x_{i,k,r}$ where $r = (s, v, \beta, p)$, according to the ordering of four-integer indices and an ordering of the number of service classes k. The antenna-beam/cell-sector and power-control values, s and p, are initialized from the estimates output by the first and second stages of the NN cascade, then optimized in the SOFM based on the residual interference levels produced by the third-stage HNN MUD. In terms of x, the demand constraints are expressed as $Dx = T$, where T is the demand array and array D consists of N subarrays of 1's and 0's. The next random call and corresponding service requirements input to the SOFM network begins a new update period of the algorithm, where a new dominant node and its neighborhood of nodes is determined and their weights modified. This procedure is repeated until the weights stabilize to a feasible 0-1 solution that is a local minimum of the optimal RRA problem. As the algorithm converges, the magnitude of weight modifications and the size of the neighborhoods are decreased. Initially, the size of the neighborhood for each subarray of W, given by $\eta = (\eta_1, \eta_2, \cdots, \eta_N)$ is large, but is decreased incrementally until $\eta_j = \|t_j\|$, the total level of service demands at BS j, for all N stations. Since the weight modifications depend on the order in which the calls are input, the SOFM approach is inherently stochastic. The SOFM must be run repeatedly to arrive at different local minima.

The following SOFM algorithm can be applied to the SRRA problem in W-CDMA networks.

1. Initialize the weight vectors of the network as $w_{j,r} = t_j/R$, which gives an initial feasible, possibly non-integer solution.

2. Randomly select a new call (with service demand k) for a BS. Represent this requirement as the input array x. Find the position j^* (BS coverage area) which is active, i.e., $x_{j^*,k} = 1$.

3. Calculate the *quality or cost potential* V_{r,j^*} for each index r in the output layer array according to (18).

4. Determine the dominant node, m_0, by competition such that $V_{m_0,j^*} = \min V_{r,j^*}, \forall r \in \mathfrak{R}$, and identify its neighboring nodes $m_1, m_2, \cdots, m_{\eta_{j^*}}$, where $\eta_{j^*} \geq \|t_{j^*}\|$ is the size of the neighborhood for input resource requirement at j^* for service class k.

5. Update the weights in neighborhood of dominant node according to the rule

$$\Delta w_{j^*,r} = \alpha(\eta, n)[e - w_{j^*,r}] \quad \forall r \ni V_{r,j^*} < V_{m_{\eta_{j^*}},j^*} \text{ where}$$

$$\alpha(\eta, n) = \frac{\alpha(n)\gamma_{j^*}}{\|t_{j^*}\|} \exp\left[\frac{-\left|V_{m_0,j^*} - V_{r,j^*}\right|}{\sigma(n)}\right],$$

which is a modified version of Kohonen's SOFM *slow* updating rule, where α and σ are positive, monotonically decreasing functions of sampled time, γ is a normalized weighting vector used in tie-breaking for a network node. For all other weight vectors, outside the neighborhood being updated, $\Delta w_{j\bullet,k} = 0$. The weights are updated as $w_{j,r} \leftarrow w_{j,r} + \Delta w_{j,r}$.

6. A hill-climbing HNN is applied to return the weight array to a feasible solution. The array w is modified around the weight adaptations of the SOFM algorithm so that $Dw = T$.

7. Repeat Step 2 until RR requirements in all cells have been selected as input vectors to the SOFM network. This forms one period of the algorithm. The procedure is repeated for K periods. In each subsequent period, α and σ are decreased according to any monotonically decreasing function.

8. Repeat Step 2 until $\|\Delta w_{r,j}\| \cong 0, \forall r, j$, this condition is considered stable convergence of the weights for a given neighborhood size. Decrease the neighborhood sizes η_j linearly for all j.

9. Repeat Step 8 until $\eta_j = \|t_j\|$, for each BS coverage area j, $j = 1, \cdots, N$.

For the multiservice demand array T, the normalized weighting vector γ, a heuristic used to damp oscillations in the algorithm updates is modified from the form used in [4] as follows

$$\gamma_i = \left| \sum_{j=1}^{N} \left(\sum_{k=1}^{K_i} t_{ik}^n c_{ij} + \sum_{l=1}^{L_i} t_{il}^p c_{ij} \right) \right| - c_{ii} , i = 1, ..., N. \tag{22}$$

Each element in γ is then normalized. The SOFM parameters can be selected heuristically, with $K = 10$, [24]

$$\alpha(0) = \min_{1 \le i \le N, 1 \le k \le K_i, 1 \le l \le L_i} (t_{ik}^n \cdot t_{il}^p), \ \alpha(n + 1) = 0.9\alpha(n),$$

$$\sigma(0) = 9, \ \sigma(n + 1) = 0.9\sigma(n),$$

$$\eta_j(0) = \|t_j\| + \lfloor N/K \rfloor, \ \eta_j(n + 1) = \eta_j(n) - 1.$$

The average probability of new call blocking, average probability of dropped handoffs and the total network capacity for each service class can be used alternately to evaluate the NN cascade as a SRRA algorithm in 3G networks. The evaluation criteria can be augmented with the terms $\kappa_j \cdot r$, representing infrastructure costs of using RRA r in coverage area j.

The initial state of the DRRA problem is the stable SRRA solution, where a new call or handoff with multiservice demands cannot be assigned to a BS without a rearrangement of existing RRAs. A time-varying traffic demand array $T(n)$ and the resource constraint relations are satisfied when $D(n)x(n) = T(n)$ at sample time n. Each epoch n represents the arrival of a single or multiple new calls to or handoffs between the cells of the network. During each period, input vectors, corresponding to the cells in which a call is placed or handoff requested, are presented to the SOFM at a rate determined by the distribution of demands in the network at that time. Since feasibility is always restored during the second stage of the SOFM, any rearrangement of the existing calls to enable a new call is automatic. If no rearrangement is possible, either the SOFM cannot converge to a feasible set of RRAs, or a feasible rearrangement may be found by allowing interference levels to increase above acceptable QoS levels. In either outcome, the call can be blocked and the previous state of the system reinstated.

For more robust convergence, step 6 of the algorithm uses Abe's approach [25] to ensure that a HC-HNN only leads to feasible RRAs that are stable points of the system of update weights. A piecewise-linear saturation function replaces the exponential used in the weight update rule in step 5. In step 7, faster updating is accomplished based on Abe's convergence acceleration for HC-HNNs to optimize integration step sizes, now applied in $K = 1$ period.[24] After the SRRA is completed to initiate the DRRA algorithm, step 8 is omitted. At the start of each epoch in the DRRA SOFM, the neighborhood function is initially set to the row vector of greatest length in the multiservice demand array $T(n)$ at sample time n.

5. SIMULATION RESULTS FOR MULTISERVICE RRA PROBLEMS

The performance of the NN cascade for RRA in W-CDMA networks is evaluated, based on simulations of multimedia extensions of cellular network models considered earlier by Kunz.[4] The interference matrices and traffic demand vector for a 25-cell network are used to represent a wireless multimedia network with non-homogenous traffic, by decomposing the number of calls at BS j in the demand vector into a row of the service demands in those calls from each real-time and packet-data subclass. Thus, the traffic demand vector becomes a multiservice traffic demand array. Time-varying traffic loading is approximated through cyclic rotation of the rows of the arrays T or periodic replacement of a selected row with a new traffic vector during the simulation run.

The computation time for the W-CDMA RRA simulations grows rapidly with the number of service classes and the number of possible RR vector selections. Simulation models are thus limited to four real-time service classes: digitized voice encoded with 8-kbps, 13-kbps, and 32-kbps codecs, as well as 64-kbps compressed video; and four packet-data service classes with rates of 64 kbps, 128 kbps, 384 kbps and 768 kbps. The RRs form a finite set of vectors. The first entry in each vector is the selection of adaptive-antenna array sectorization (beamforming) as any number of omni-, 180°-, 120°-, 45°-, 72°-, or 60°-sectors at each BS, such that the total of the sectors equals 360°. The second entry is the selection of activity monitoring with 0 for "off" and 1 for "on". The third entry is the selection of spreading factor, based on 1.92-Mcps and 3.84-Mcps spread channels, of 4, 16, 64 and 256, with rate matching assumed to align information rates with chip rates. The fourth entry is the selection of power control at the mobiles, with 0-level or no power control, 4-level control, 20-level control, 40-level control over a 10-dB reference transmit power range. Each resource vector $r = (s, v, \beta, p)$ is mapped to an estimated interference value I_r based on MAI statistics collected in microcellular networks and the minimum SIR values equivalent to the QoS required by service classes active in the row corresponding to each BS in array T. The interference establishes actual CDMA frequency reuse performance, and thus determines the number of common and dedicated channels available to meet multiservice demands. The cost coefficient vector for each assignment r is $c = (1, 1, 1, 1)$. Individual cost terms $r \cdot c$ are summed over the number of active BSs in the network and the number active service classes at those stations and added to the objective function.

In order to exercise the three NN stages of channel estimation, antenna array control, and MUD, additional network features are set. The BS transceivers are assumed to use rate 1/3, constraint length $k = 9$ convolutional encoders. The multipath delay model is a 2- to 8-path profile, with the principal paths ordered by magnitude according to the recommended IMT-2000 channel propagation model. Each path is subjected to independent Rayleigh fading with power scaled to the IMT-2000 model and Doppler frequency of $f_D = 100$ Hz. The BER performance for the HNN MUD has been measured in laboratory experiments as a function of the average E_b/N_0, as the number of active users is varied. Average BER monotonically falls as average E_b/N_0 increases, while that of the CD receiver approaches an error floor that depends on K the number of active users. As K increases, the E_b/N_0 loss from the single-user case increases due to residual MAI. When $K = 8$, however, E_b/N_0 loss at BER = 10^{-3} is only about 2.5–3 dB. The operation of the RNN antenna array control (AAC) is unlike a fixed multibeam antenna, considered earlier.[4] The AAC can change beam direction finely, initiates coverage from the omnibeam pattern. It forms the optimum beam pattern adaptively and can direct the beam toward the resolved path of each user to realize coherent Rake combining even though their arrival angles are quite different. Values used for the RNN AAC are based on experiments with a channel-fading simulator, with K users, one desired and $K - 1$ interfering users. For a comparison of RNN AAC and two-antenna diversity reception, average BERs of the AAC were measured as a function of received power ratio of interfering users to desired user. The data rate and chip rate are 64 kbps and 4.096 Mcps, respectively. Simulated paths of all mobile users arrive from the same direction. The average E_b/N_0 and the arrival angle of the desired user are set to 11.7 dB and 62°, respectively. Average BER reduction is about one order of magnitude compared to the antenna diversity case, even if the worst-case interferer's power is 10 dB higher, corresponding to a single user with 10 higher data rate. The improvement offered by the RNN AAC diminishes as the arrival angle of the interferers' signals approach that of the desired user's signal. When two users are within the beamwidth, either one could be blocked. These empirical results are used in the network simulations.

Simulations were performed on a PC, based on adaptive learning models included in MATLAB's Neural Network Toolbox, with custom C routines to implement network models as well as the MLP-NN,

RNN, HNN and SOFM of the NN cascade. Due to run-time limitations, once presented with $T(n)$ at each iteration of the simulation, the four NN stages of RRA algorithm are run sequentially, with outputs of the preceding stage used as inputs to succeeding stages. Algorithm performance is measured alternately on the basis of average probability of call blocking, average probability of dropped handoffs, and the total active service classes in the network (channel capacity), together with the average number of iterations (ANIs) required for asymptotic convergence, based on a prescribed error value.

The traffic demand vector $[10, 11, 9, 5, 9, 4, 5, 7, 4, 8, 8, 9, 10, 7, 7, 6, 4, 5, 5, 7, 6, 4, 5, 7, 5]^T$, introduced by Kunz for a voice-only network of 25 base stations, is expanded to the following 8×25 array of multiservice demands. For simplicity, each call is assumed to require only one class of service.

$$T = \begin{bmatrix} 3\ 3\ 00\ 01\ 0\ 3\ 02\ 01\ 5\ 0\ 30\ 00\ 00\ 10\ 00\ 0 \\ 2\ 2\ 20\ 21\ 0\ 2\ 02\ 01\ 5\ 0\ 20\ 02\ 0\ 2\ 14\ 02\ 0 \\ 0\ 1\ 20\ 31\ 0\ 2\ 14\ 01\ 0\ 0\ 26\ 03\ 0\ 2\ 10\ 02\ 0 \\ 1\ 1\ 11\ 01\ 1\ 0\ 00\ 01\ 1\ 1\ 00\ 20\ 0\ 3\ 10\ 10\ 0 \\ 2\ 0\ 11\ 20\ 1\ 0\ 10\ 21\ 2\ 2\ 00\ 20\ 2\ 0\ 10\ 13\ 0 \\ 0\ 0\ 11\ 00\ 1\ 0\ 10\ 22\ 0\ 1\ 00\ 00\ 2\ 0\ 10\ 10\ 3 \\ 2\ 2\ 11\ 20\ 2\ 0\ 10\ 21\ 2\ 2\ 00\ 00\ 1\ 0\ 00\ 10\ 2 \\ 0\ 2\ 11\ 00\ 0\ 0\ 00\ 20\ 0\ 1\ 00\ 00\ 0\ 0\ 00\ 10\ 0 \end{bmatrix}^T \quad \begin{matrix} 8- \text{kbps voice} \\ 13- \text{kbps voice} \\ 32- \text{kbps voice} \\ 64- \text{kbps video} \\ 64 - \text{kbps packet data} \\ 128 - \text{kbps packet data} \\ 384 - \text{kbps packet data} \\ 768 - \text{kbps packet data} \end{matrix}$$

The same 25×25 interference matrix introduced for Kunz' Helsinki model is used in the simulations.

The SOFM for the SRRA problem for this network is simulated, with $K = 100$ initial states and RRAs all initialized to (1, 0, 64, 0) in each BS coverage area. The average combined blocking and dropped handoff probability is 0.087 for the real-time classes and 0.173 for the packet-data classes, with the ANI equal to 1874.2 for the final SOFM stage. Even with an eight-fold increase in the complexity of the multimedia network over Kunz' original model, convergence of NN-cascade RRA algorithm occurs in fewer iterations than the convergence of HNN algorithm in 2450 iterations reported by Kunz for his voice network. The cascade simulation is very slow due to the sequential operation of the stages in generating estimates. The SOFM is slower than the single-service SOFM evaluated in [4], since it must "learn" the correct RRA iteratively over a larger search space to meet an array of demands. The algorithm produces higher blocking probabilities for packet-data demands, since only a limited number of DCHs can be allocated to these classes, while real-time services have access to all DCHs supported by the RRA. The algorithm slowly increments sectorization values to 6, sets activity monitoring "on", while power control assignments vary over the 25 BS areas according to the number of active high-rate packet data requests.

To evaluate the SOFM for DRRA, the rows of T above are cyclically shifted 5 positions down every Π periods, with $\Pi = 10, 20, 50$, and 100, to represent dynamic local demands at BSs. In order to examine the algorithm sensitivity to initial RRAs, as it responds to demand shifts, three patterns in each BS area are initially used in the simulation: (1) RRAs are set to (1, 0, 64, 0); (2) RRAs are set to (3, 1, 64, 20);

and (3) RRAs are set to the final values of the SRRA after $K = 100$ periods. In response to these cyclic demand shifts, the average blocking/dropped handoff probabilities for the DRRA with initial RRA pattern (1) increase from 0.087 to 0.291 for real-time services and from 0.173 to 0.426 for the packet-data services, as the number of periods Π decreases from 100 to 10, respectively. For initial RRA pattern (2), the average blocking/dropped handoff probabilities increase from 0.057 to 0.206 for real-time services and from 0.098 to 0.323 for packet-data services, as Π decreases from 100 to 10, respectively. Lastly, using the final RRA patterns from the SRRA problem results in the average blocking/dropped handoff probabilities from 0.034 to 0.157 for real-time services and from 0.078 to 0.299 for packet-data services, as Π decreases from 100 to 10, respectively.

6. CONCLUSIONS

A cascade model of an MLP-NN for channel estimation, an RNN for adaptive antenna control, a discrete-form HNN for joint multiuser detection, and a discrete-space Kohonen SOFM has been proposed for the problem of allocating RRs to meet QoS requirements of multimedia service demands in 3G wireless networks. W-CDMA network parameters on uplinks have been assumed to model the resources available to support the diverse SIR and delay requirements for variable-rate audio, high-rate packet data, and real-time video. Simulation results for each of the first three NN stages have been presented for representative W-CDMA scenarios. Finally, both static and dynamic versions of the complete NN cascade algorithm have been simulated for the RRA of multimedia extensions of published cellular network models. The simulation results have been informally compared to earlier published results for single-stage HNN and SOFM techniques applied to resource allocation in single-service voice and data networks.

ACKNOWLEDGMENTS

The author wishes to thank his colleagues at FIT for their support and to commend the ITU for its leadership in promoting universal standards for next-generation wireless mobile networks.

REFERENCES

1. T. Ojanperä and R. Prasad, "An overview of air interface multiple access for IMT-2000/UMTS," *IEEE Commun. Mag.*, vol. 36, no. 9, pp. 82–95, Sept. 1998.
2. E. Dahlman, B. Gudmundson, M. Nilsson, and J. Sköld, "UMTS/IMT-2000 based on wideband CDMA," *IEEE Commun. Mag.*, vol. 36, no. 9, pp. 82–95, Sept. 1998.
3. K. Das and S. D. Morgera, "Interference and SIR in integrated voice/data wireless DS-CDMA networks – a simulation study," *IEEE J. Select. Areas Commun.*, vol. 15, pp. 1527–1538, Oct. 1997.
4. W. Hortos, "Self-organizing feature maps for dynamic control of radio resources in CDMA microcellular networks," *Appl. and Sci. of Artificial Neur. Net. II, Proc. of SPIE*, vol. 3390, Orlando, FL, pp. 378-391, Apr. 1998.

5. K. Gilhousen, I. Jacobs, R. Padovani, A. Viterbi, L. Weaver, and C. Wheatley, "On the capacity of a cellular CDMA system," *IEEE Trans. Veh. Technol.,* vol.50, no. 2., pp. 303-312, 1991.

6. G. L. Turin, "The effects of multipath and fading on the performance of direct-sequence CDMA systems," *IEEE J. Select. Areas Commun.,* vol. SAC-2, pp. 507–603, July 1984.

7. Jalali and P. Mermelstein, "On the bandwidth efficiency of CDMA systems," in *Proc. IEEE ICC '94,* May 1994, pp. 515-519.

8. Y. C. Yoon, et,.al., "A spread-spectrum multiaccess system with cochannel interference cancellation for multipath fading channels," *IEEE J. Sel. Areas of Commun.,* vol. SAC-11, No. 7, pp. 1067-1075, Sept. 1993.

9. R. Fraile and N. Cardona, "Fast neural network method for propagation loss prediction in urban environments," *Electron. Lett.,* vol. 33, no. 24, pp. 2056–2058, 1997.

10. J. Walfisch and H. L. Bertoni, "A theoretical model of UHF propagation in urban environments, " *IEEE. Trans. Antennas Propag.,* vol. 36, no. 12, pp. 1788 –1796, 1988.

11. S. R. Saunders and F. R. Bonar, "Explicit multiple building diffraction attenuation function for mobile radiowave propagation," *Electron. Lett.,* vol. 27, no. 14, pp. 1276–1277, 1991.

12. E. Gschwendtner and F. M. Landstorfer, "Adaptive propagation modelling using a hybrid neural network technique," *Electron. Lett.,* vol. 32, no. 3, pp. 162–164, 1996.

13. *EUROCOST: European cooperation in the field of scientific and technical research. COST 231, Final Report,* 1996.

14. M. Benson and R. A. Carrasco, "Recurrent neural network array for CDMA mobile communication systems," *Electron. Lett.,* vol. 33, no. 25, pp. 2105–2106, 1997.

15. R. J. Williams and D. Zipser, "A learning algorithm for continually running fully recurrent neural networks," *Neural* Comput., vol. 1, pp. 270-280, 1989.

16. S. Verdu, "Minimum probability of error for asynchronous Gaussian multiple-access channels," *IEEE Trans. on Info. Theory,* vol. 32, pp. 85–86, Jan. 1986.

17. U. Mitra and H. V. Poor, "Adaptive receiver algorithms for near-far resistant CDMA," *IEEE Trans. Commun.,* vol. 43, pp. 1713–1724, 1995.

18. G. I. Kechriotis and E. S. Manolakos; "Hopfield neural network implementation of the optimal CDMA multiuser detector," *IEEE Trans. on Neural Networks,* vol. 7, pp. 131–141, Jan. 1996.

19. G. I. Kechroitis and E. S. Manolakos, "A hybrid digital computer – Hopfield neural network spread-spectrum CDMA detector for real-time multi-user demodulation," *Proc. 1994 IEEE-SP Int. Workshop on Neur. Networks for Signal Processing,* Ermoni, Greece, pp. 545–554, Sept. 1994.

20. S. Verdu, "Computational complexity of optimum multiuser detection," *Algorithmica,* vol. 4, pp. 303–312, 1989.

21. M. K. Varanasi and B. Aazhang, "Multistage detection in asynchronous code-division multiple access communications," *IEEE Trans. on Comm.,* vol. 38, pp. 509–519, Apr. 1990.

22. J. J. Hopfield and D. W. Tank, " Neural computation of decisions in optimization problems," *Biological Cybern.,* vol. 52, pp. 141-152, 1985.

23. T. Kohonen, "Self-organized formation of topologically correct feature maps," *Bio. Cybern.,* vol. 43, no. 1, pp. 59-69, 1982.

24. K. Smith and M. Palaniswami, "Static and dynamic channel assignment using neural networks," *IEEE J. Selected Areas Comm.,* vol. 15, no. 2, pp. 238-249, 1997.

25. S. Abe, "Global convergence and suppression of spurious states of the Hopfield neural nets," *IEEE Trans. Circuits Syst.,* vol. CAS-40, no. 4, pp. 246-257, 1993.

Successive Interference Cancellation for Interception of the Forward Channel of Cellular CDMA Communications

Michael Golanbari and Gary E. Ford
Center for Image Processing and Integrated Computing and
Dept. of Electrical and Computer Engineering
University of California, Davis, CA 95616
{golanbar,ford}@ece.ucdavis.edu

Abstract

We develop and evaluate receiver signal processing algorithms for the detection of signals transmitted via the forward link of a cell in a cellular system modeled after the IS-95 standard for direct-sequence spread-spectrum code-division multiple-access (CDMA) communications. Multiuser detectors on board airborne and terrestrial mobile interceptors or monitors attempt the simultaneous detection, in a single receiver, of all communication signals transmitted by the base station of interest. Due to the detrimental effects of transmitter, receiver and channel nonlinearities, very fast multipath fading, shadowing, path loss, Doppler spread, additive white Gaussian noise, and intracell and intercell multiple-access interference, the user signals are de-orthogonalized. This leads to performance degradation in conventional receivers that is too severe, especially when the powers of some of the interfering users are dominant. In order to improve upon the performance of conventional matched filter receivers, this article focuses on the development and evaluation of fast and reliable successive interference canceling (SIC) algorithms. The techniques we have developed can be used to enable successful interception of CDMA signals; to relax the strict requirements on power control; and to improve the capacity of CDMA systems.

1 Introduction

To provide communication services to a large and growing number of users, CDMA cellular systems reuse the same frequencies within geographical cells. To further increase the capacity of cellular communications, new receiver structures are required to provide mitigation from the harmful effects of the resulting co-channel interference. The objective of the research reported in this article is the development and evaluation of reliable receiver signal processing algorithms to jointly demodulate the signals employed in the forward link of a single cell in a system modeled after the IS-95 standard for CDMA cellular communications. The signals are received at a single sensor on board an airborne or satellite interceptor/monitor and a terrestrial interceptor/monitor in the presence of transmitter, receiver and channel nonlinearities, very fast time varying multipath fading, shadowing, path loss, Doppler spread, intra-cell interference, inter-cell interference and additive white Gaussian noise (AWGN). The effects of nonlinearities and fading are independent from chip to chip, effectively de-orthogonalizing the transmitted signals. A multiuser detector on board the airborne and terrestrial interceptors attempts the simultaneous detection of all communication signals in a single receiver.

The techniques we have developed can be used to enable successful surveillance and/or reconnaissance of CDMA signals for law enforcement, defense, cellular fraud management, etc.; to improve the capacity of existing and proposed ground to air and sea to air (maritime) communications systems, such as satellite and aeronautical communications platforms; and to potentially relax the stringent requirements on power control imposed by the IS-95 system.

The research is important for several reasons:

- The high deployment rate of new IS-95 cellular CDMA systems in the US and abroad, and the emergence of CDMA as a strong candidate for the air interface of the universal personal communications network planned for the near future, necessitate the design and implementation of practical, interference-resilient demodulators for co-channel spread-spectrum signals.

- The stringent requirements on power control imposed by the IS-95 system to combat the near-far problem may be relaxed if multiuser detection is employed. This work exploits this observation and proposes to apply multiuser detection to the signals transmitted on the downlink of IS-95. Power control dictates significant reductions in the transmitted powers of the strong users in order for the weaker users to achieve reliable communication. This can become self defeating since it can actually decrease the overall multiple access and anti jamming capabilities of the system. Before the emergence of solutions to the near-far problem based on multiuser detection, the only remedies available were power control and design of signals with ever more stringent cross correlation. Thus, solution to the near-far problem has been highlighted as an important task of multiuser detection.

- We believe that the problem is new. Although the general problem of signal interception has received some attention in the literature [1], and even though several multiuser detection schemes have been previously applied to CDMA signals [2–5], to our knowledge very little has been published on the interception of multiuser IS-95 signals. The few exceptions include [6,7], which did not consider SIC nor airborne interception. References [3–5] have considered successive and parallel interference cancellation with convolutional forward error correction coding and decoding, showing improved performance compared to the use of conventional matched filter receivers. Reference [8] has demonstrated performance gains for multiuser detection on the forward link of a CDMA system, but it did not consider SIC. Furthermore, most of the previously published performance results, including [2–7], do not consider the non-linear, very fast fading channel models which we investigate here, and they concentrate on the reverse link of CDMA communications. These results are not fully applicable to the scenario of signaling on the forward channel of IS-95 because the transmitter and the propagation channels which we investigate for the forward link of IS-95 are substantially different than those which have been previously considered in the literature for the reverse link. The problem offers some new twists – e.g., the particular structure of IS-95 signals, the fact that from the point of view of the interceptor, power control increases the dynamic range and worsens the near-far effect, and the fact that we are interested in intercepting the signals of **all** the users in a desired cell, rather than only a single user. Numerous articles, such as [8–11] reported on the relatively unsatisfactory performance of some conventional matched filter receivers in very harsh downlink propagation channels without multiuser detection and/or antenna diversity. This motivates the need to investigate multiuser receivers which are specifically designed for more practical and realistic models of the downlink of the IS-95 system and for the transmitter, channel and receiver nonlinearities and severe fading conditions which may be associated with the downlink under certain circumstances.

The paper is organized as follows. In Section 2, we present the problem which is addressed in this article. Section 3 provides the models for the signal and the interference canceling receivers. Section 4 contains performance results generated from computer simulations, comparing the interference canceling detectors to the non-interference canceling detector and the single user bound. Finally, in Section 5, we provide some concluding remarks and issues to be researched in the future.

2 Statement of the Problem

The problem to be addressed is the following: given the received signal, a sum of K co-channel direct-sequence spread-spectrum signals in noise and interference, we wish to develop algorithms to simultaneously detect the digital messages x_{mk}, for all bit time instances m and all users $k = 1, 2, \ldots, K$. Using computer simulations, we have determined the bit-error-rate (BER) performance as a function of signal powers and number of users in the desired cell. Specifically, we have designed and implemented in software two successive interference cancellation schemes. Successive cancellation [2–4] is based on demodulating the strongest user using conventional methods, and remodulating the recovered message. The remodulated signal is then subtracted from the received composite signal, leaving an approximation to the sum of signals due to the remaining users. This process is then repeated, and each time the signal due to the strongest remaining user is subtracted, resulting in a waveform with substantially diminished interference. The disadvantages of the technique are it's suboptimal performance, it's requirements that the received amplitudes and phases be estimated with good accuracy, and that some power separation exist between the strongest signal and the next-strongest signal in each step of each successive cancellation stage; otherwise, its performance degrades.

Successive interference cancellation schemes are derived in [2–4], demonstrating significant performance improvements over the conventional receiver which does not employ interference cancellation. We have simulated in software a conventional matched filter receiver (CMF), a conventional multistage successive interference canceling receiver (CSIC) and a modified multistage successive interference cancellation (MSIC) scheme which employ coherent detection and pilot signals to obtain channel estimates. The results show that the modified successive interference canceler provides traffic capacity increase over the capacity of the conventional successive interference canceling receiver, which in turn provides capacity increase over the conventional matched filter receiver.

3 Signal Model and Interference Cancellation Receivers

The signal model is based on the IS-95 CDMA cellular system, which applies a universal one-cell frequency reuse: On the base station to mobile link, signals are transmitted over a common portion of the frequency spectrum. Each cell has a common pilot channel which is transmitted at all times by the base station on each active forward CDMA channel. The user signals are orthogonalized, as all signals emanating from the same base station transmitter are synchronized. When the propagation path between the base station of interest and the interceptor is an AWGN channel, the traffic channels are orthogonal and synchronous and multiuser detection is not necessary on the forward link, as a bank of simple correlation receivers (ie., the CMF receiver) is optimal.

We consider channel models in which this signal orthogonalization is not preserved. The communications between the base station and the airborne interceptor are assumed to take place over a Rician flat-fading mobile satellite channel, while the communications between the base station and a land-based mobile interceptor are assumed to take place over a Rayleigh frequency-selective fading mobile channel. For both channels, we assume very fast time-varying fading which is independent from chip to chip. We also incorporate, in all our simulations, the effects of transmitter, receiver and channel nonlinearities, all of which effectively destroy the orthogonality of the traffic channels on the forward link. Both channels introduce multipath interference, log-normal shadowing, path loss, Doppler spread, intracell interference, intercell interference and additive white Gaussian noise.

Through power control in the downlink, the power transmitted to close-in portables is reduced, while the signal to interference requirements of all portables are satisfied, increasing the overall capacity [9]. In our signal model, we consider power control models designed to follow the slow variations in received signal to interference ratio due to shadow fading and path loss (slow power control), as well as the fast variations due to fast multipath fading.

From the point of view of the airborne or land-based (terrestrial) interceptor, power control on the forward link can be a major problem, because the higher power allocated by the base station to transmit to mobiles which are further away from the base station can overwhelm the power transmitted by the base to mobiles which are close to the base. Hence, *from the point of view of the interceptor*, this particular power control scheme is not beneficial if the interceptor employs a conventional matched filter receiver which is not near-far resistant. However, as we show in the sequel, the power control scheme is beneficial if the interceptor employs successive cancellation, because the successive interference canceling detectors perform better when the signals are of distinctly different powers [2].

Note that in this paper, we initially address only the two channel models mentioned above. This limits the scope of our work. The baseband models that we have developed in software for the forward traffic propagation channels are based on the multipath models described in [12–14]. For frequency-flat Rician fading (satellite or airborne interceptor), the channel parameters are set as in [12], and there is a direct line of sight path between the desired base station transmitter and the receiver; for frequency-selective Rayleigh fading (terrestrial interceptor), the delays between the received replicas of the transmitted signal, τ_c, are random integer multiples of the chip duration T_c, there is no direct line of sight path between the base station transmitter and the receiver, and the channel parameters are set according to [13,14]. The total power transmitted by the base station is normalized to unity, with 20% of the power allocated to the pilot signal, and a fraction of the remainder of the power allocated for communicating with each portable which is in contact with the base station of interest.

To operate in the cochannel interference signal environment described above, we have implemented the conventional matched filter receiver and the conventional and modified multistage successive interference canceling receivers in software. The SIC receivers are shown in block diagram form in Figs. 1, 2, and 3. The conventional K-user demodulator commonly employed in practice is implemented as a bank of optimum detectors for single-user communications. There is one matched filter or RAKE receiver for each leg of the quadrature demodulator for each user, followed by one Viterbi decoder for the convolutional code of each user. The performance of this demodulator is used as one baseline against which we compare the conventional and modified multistage successive interference canceling receivers. For the interference canceling receivers, at each step of each stage of interference cancellation, we weight the re-spread and re-constructed user signal by a partial-cancellation factor. We employ this weighting procedure in order to reduce the effects of imperfect signal reconstruction and cancellation at each step of each stage of interference cancellation. This way, more reliable estimates (ie., those corresponding to users which were received with higher powers) receive higher weight in the multiple access interference reconstruction and subsequent cancellation operations. We determined the proper weights by an optimization procedure.

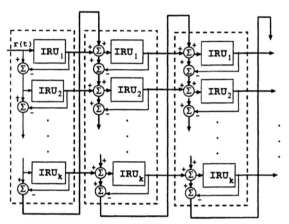

Figure 1: Multistage successive interference cancellation. The block IRU_k stands for interference regeneration unit for user $k = 1, 2, ..., K$ at each step of each stage.

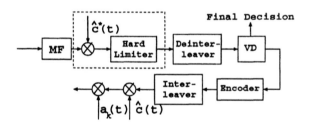

Figure 2: Details of each cancellation unit for the modified successive interference canceler (MSIC). The dashed block is used only with hard decision decoding. MF and VD stand for matched filter and Viterbi decoder, respectively. The quantity $\hat{c}(t)$ represents the channel estimate, $\hat{c}^*(t)$ it's complex conjugate, and $a_k(t)$ the spreading waveform for user k. The block diagram applies to the case of flat fading; for frequency-selective fading, there is one such cancellation unit for each finger of the RAKE receiver.

In IS-95, a block of reference symbols (the pilot sequence) is added in parallel to the data stream before transmission over the channel. The received signal is down-converted to baseband (in this project, we assume ideal carrier frequency and phase acquisition and tracking) and correlated with a locally generated replica of the known reference symbols to obtain unbiased but noisy preliminary channel estimates. The real and imaginary correlation values (obtained

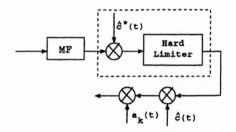

Figure 3: Details of each cancellation unit for the conventional successive interference canceler (CSIC). The dashed block is used only with hard decision decoding for a final hard decision if necessary. The block diagram applies to the case of flat fading; for frequency-selective fading, there is one such cancellation unit for each finger of the RAKE receiver.

from the I- and Q- channels, respectively) are evaluated at the sampling instants and stored in memory for the entire length of the incoming sequence. The locally generated replica of the known pilot sequence is then shifted by one chip period, and the correlation procedure is repeated. The correlation vector contains the information needed for sequence synchronization: for the case of frequency-selective fading, the index of the maximum value of the correlation vector gives the delay between the strongest incoming ray and the local pilot sequence, and the delays of the remaining trackable paths are found from successively searching for additional peaks in the correlation vector. The indices of the peaks and their magnitudes and phases are further processed using a subspace-based iterative algorithm to compensate for the delays, amplitude scalings, and phase rotations introduced by the channel, for every tracked path. The channel estimation algorithm is based upon the iterative method described in detail in [15].

In decoding the binary convolutional code employed on the downlink, we have implemented a modified branch metric for use in the Viterbi algorithm: we use the estimates of the channel gain to compute the metric by evaluating the squared Euclidean distance between the samples at the outputs of the matched filters and the candidate symbols after weighing the latter by the channel gains. This enables us to provide information on channel reliability to the soft-decision Viterbi algorithm employed at the decoder. The essence of modifying the branch metrics is the relative accentuation of more credible information and the relative suppression of less credible information. Our numerical results demonstrate that the modified branch metric leads to improvements in BER performance compared to the case of hard decision decoding when channel estimation errors are present, which is the case in practice.

4 Single Cell Performance in Multicell Environment

In this section, we describe our study of the performance of the conventional matched filter receiver and the conventional and multistage successive interference canceling receivers in the cochannel interference environment.

We have conducted extensive computer simulations to estimate receiver performance. We have simulated the different powers assigned to different traffic channels (users), modifying the power separations between users at each Monte-Carlo run to account for user mobility. Performance is reported as average bit error rate (BER) as a function of the average bit energy to noise power spectral density E_b/N_o (both quantities are averaged over all active users in the cell of interest). The simulation model is composed of 7 hexagonal cells, each with a base station at its center. The base station of interest is located in the center of this cell cluster and is comprised of three contiguous sectors, each sector occupying 120°. Transmissions from the other six base stations interfere with the transmissions from the base station of interest. The interference from base stations outside these six is assumed insignificant.

We have investigated a potential major cause of performance degradation due to the active use of the same Walsh codes by base stations in both the desired cell and adjacent cells. Our model for intercell and intracell interference incorporates power allocation to the various traffic channels, soft handoff, mobile speed, spatial decorrelation of shadowing [13], path loss, Doppler spread and multipath. We pass the interfering signals from the undesired base stations through fading channels that are independent of the fading channel between the interceptor and the desired base station in the cell of interest. We assume the interceptors are much closer to the base station of interest than they are to the undesired base stations, and hence the received intercell interfering signals are attenuated compared to the desired signal. For simplicity, we assume that the number of users in each of the interfering cells is equal to the number of users in the desired cell, and the active users in the interfering cells are all using the same exact subset of Walsh codes as those used by the active users in the desired cell. The reuse of the same subset of Walsh codes in adjacent cells should result in worst-case performance of all receivers investigated in this study.

Figures 4 – 5 show the average BER performance per user vs. the average bit energy to noise power spectral density E_b/N_o. Results are shown for the conventional matched filter (CMF), conventional successive interference canceling (CSIC) and modified successive interference canceling (MSIC) receivers operating in a single cell, flat Rician fading environment with intercell and intracell multiuser interference. The CSIC and MSIC receivers employ two stages of cancellation each. For comparison purposes, baseline performance is given for a conventional receiver employing a rate 1/2, constraint length 9 convolutional encoding and decoding when only a single user is active in a single cell with no multicell interference, in flat Rician fading with perfect channel estimation of the fading parameters (amplitude and phase), but without any estimation of the AWGN. The performance of this receiver in this signal environment is equivalent to the performance of the optimal receiver in AWGN with coding [2]. When the number of users $K = 15$, the performance of the MSIC is the closest to the single-user bound ($K = 1$) among all receivers considered in this discussion and is superior to the CSIC and the CMF for $K = 15$. As the number of users increases, the gap in performance between the MSIC receiver and the other two receivers is even more pronounced, since the MSIC lowers the BER floors associated with the competing receivers.

In general, the intercellular to intracellular interference ratio is a random variable, since the interference powers from all surrounding cells will be a function of the random numbers of users in adjacent cells, as well as random path loss exponent, shadowing, Doppler spread and voice activity. However, in our simulations we assumed that the path loss exponent for intercell interference in all simulation runs were four and three for the airborne and terrestrial interceptors, respectively. We have also simulated the effects of errors in the power control algorithm. For BER of 10^{-3}, the capacity of the system, *from the point of view of the interceptor,* is virtually unchanged compared to the case of perfect power control, assuming the receiver is capable of accurately tracking the time-varying powers of the data channels[1]. This is because successive interference cancellation algorithms perform better when the power separations between users

---

[1]However, from the point of view of the mobile users communicating with the base station of interest, the capacity is adversely affected due to errors in the power control algorithm.

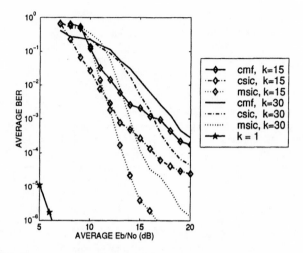

Figure 4: BER Performance of a conventional matched filter receiver, a conventional interference canceling receiver and a modified successive interference canceling receiver operating in a multi cell, frequency-flat Rician fading environment with multiuser interference, with k active users in the cell.

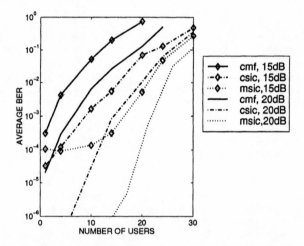

Figure 5: Capacity of a conventional matched filter receiver, a conventional interference canceling receiver and a modified successive interference canceling receiver operating in a multi cell, frequency-flat Rician fading environment with multiuser interference. The legend also indicates the average bit energy to noise power spectral density per user.

are more distinct [2]. This compares favorably with the capacity of the same system employing the conventional matched filter receiver and the conventional interference canceler.

Figs. 6 - 7 depict the performance of the two-stage successive interference canceling re-

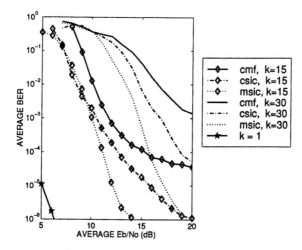

Figure 6: Performance of a conventional matched filter receiver, a conventional interference canceling receiver and a modified successive interference canceling receiver operating in a multi cell, frequency-selective Rayleigh fading environment with multiuser interference. The letter k denotes the number of active users in the cell.

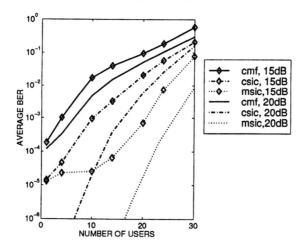

Figure 7: Capacity of a conventional matched filter receiver, a conventional interference canceling receiver and a modified successive interference canceling receiver operating in a multi cell, frequency-selective Rayleigh fading environment with multiuser interference.

ceivers with perfect power control, hexagonal cell geometry and path loss exponent of three on a frequency-selective Rayleigh fading channel. The channel model consists of two independent paths. The delays between the paths are assumed to be random integer multiples of the chip period T_c. Both paths are assumed to be tracked by a RAKE receiver utilizing the pilot sequence for channel estimation as described above. All receivers employ one matched filter for each user on each finger of the RAKE receiver; the outputs of the RAKE fingers are combined via equal-gain combining to yield a decision statistic that is used for data symbol estimation. For the successive interference cancellation schemes, there is one multistage successive interference canceler for each finger. The assumed path loss exponent for the intercell interference for the terrestrial interceptor is three. All receivers are operating in a multicell scenario with intercell and intracell multiuser interference (with the exception of the baseline receiver ($K = 1$) which does not suffer from multiuser interference.) When the number of users $K = 15$, the performance of the MSIC is the closest to the single-user bound ($K =1$) among all receivers considered in the figure, and superior to the CSIC and the CMF. As the number of users increases, the gap in performance between the MSIC receiver and the other two receivers is even more pronounced, since the MSIC lowers the BER floors associated with the competing receivers. For BER of 10^{-3}, the capacity of the system, from the point of view of the interceptor, degrades only slightly compared to the case of perfect power control, assuming the receiver can accurately track the data channel powers in the presence of power control errors.

The capacity of the system in a multicell scenario, like that for a single cell scenario, is slightly smaller with the flat Rician-fading channel than it is with the frequency-selective Rayleigh fading channel when the number of users is not too large. This is because the RAKE receiver employed in the frequency-selective channel provides additional processing gain by combining the outputs of the RAKE fingers dedicated to the signals from the two independently fading paths. However, the capacity in a frequency-flat fading channel is slightly better than that of a frequency-selective channel when the number of users is relatively large or when the signal to noise ratio is small, because then the signal at each RAKE finger suffers from too much interference from the signal that has propagated along the other path. In multicell scenarios, the detectors exhibit BER floors, due to the additional interference and the accumulated errors from imperfect regeneration and cancellation of other users.

5 Summary

We have employed successive interference cancellation techniques to simultaneously detect multiple cochannel signals transmitted on the forward link of the IS-95 CDMA cellular system. The signals are received at airborne and terrestrial mobile interceptors in the presence of transmitter, channel and receiver nonlinearities, very fast (chip to chip) time varying frequency-flat Rician fading and frequency-selective Rayleigh fading, shadowing, path loss, Doppler spread, additive white Gaussian noise, intracell interference and intercell interference. These transmitter, channel and receiver impairments act to severely damage the orthogonality of the traffic channels, necessitating multiuser detection. The forward channel employs power control, creating a near-far problem at the interceptor. The reference pilot channel was exploited for channel estimation and coherent detection as well as interference cancellation. We have shown that the reference pilot-channel assisted coherent multistage successive interference canceling receiver, which uses the decisions from the output of the error control decoder for signal regeneration at each step of each stage of interference cancellation, performs the closest to the coherent receiver which is using perfect channel estimates (ideal coherent receiver), providing capacity gains over the other detector implementations which were considered here.

The main drawbacks of the interference cancellation techniques discussed in this presentation are suboptimal performance, the need for accurate estimates of received signal amplitudes and chip, bit and frame timings, accurate estimates of carrier phase and frequency and signature codes of all desired users, some power separation between the strongest traffic channel and the next-strongest traffic channel in each successive cancellation stage[2], and processing delay. These parameter estimates are not easy to obtain in practice. Most of the required parameters can be estimated from the pilot channel, and the required signature codes can be estimated by using one of a number methods proposed in the literature for code waveform estimation, such as [16], for example. The requirements of minimum delay and implementation simplicity necessitate the need to limit the number of cancellations. However, we have shown that with careful design of system parameters, such as user signature code waveforms with very low cross-correlation properties, accurate power control, powerful forward error correcting channel codes and other system parameters, these interference cancellation methods can provide satisfactory performance, which tracks the performance of the optimum receiver, and they are near-far resistant. Furthermore, in general they are substantially easier to implement than the optimal receiver: successive cancellation requires computational complexity per symbol which is *linear* in the number of users K, in contrast to the optimum demodulator, which has complexity per symbol that is *exponential* in K.

In the near future, we intend to investigate the robustness of the receivers to parameter estimation errors and the application of antenna arrays combined with interference cancellation to the problem of signal interception.

References

[1] W. A. Gardner, "Signal interception: a unifying theoretical framework for feature detection," *IEEE Trans. Comm.*, vol. 36, (no.8), pp. 897–906, Aug. 1988.

[2] S. Verdu, "Multiuser detection," *New York: Cambridge University Press*, 1998.

[3] M. R. Koohrangpour and A. Svensson, "Joint interference cancellation and Viterbi decoding in DS-CDMA," *Proc. IEEE PIMRC Conf.*, vol. 3, pp. 1161–5, Sept. 1997.

[4] M. Brandt-Pearce and M. H. Yang, "Soft-decision multiuser detector for coded CDMA systems," *Proc. IEEE Int'l. Conf. Comm.*, pp. 365–9, Atlanta, Georgia, June 1998.

[5] Y. Sanada and Q. Wang, "A co-channel interference cancellation technique using orthogonal convolutional codes on multipath Rayleigh fading channels," *IEEE Trans. Veh. Tech.*, vol. 46, No. 1, pp. 114–128, Feb. 1997.

[6] A. McKellips and S. Verdu, "Multiuser detection for eavesdropping in cellular CDMA," *Thirty-First Asilomar Conference on Signals, Systems and Computers*, vol. 2, pp. 1395–9, Nov. 1997.

[7] A. McKellips and S. Verdu, "Eavesdropping syndicates in cellular communications," *Proc. IEEE VTC*, vol. 1, pp. 318–22, May 1998.

[8] A. Klein, "Data detection algorithms specially designed for the downlink of CDMA mobile radio systems," *1997 IEEE 47th VTC*, vol. 1, pp. 203–7, May 1997.

---

[2]In practice, the power separation which is needed is at least 1 dB. In this contribution, we consider power separations of 0.5 to 6 dB.

[9] A. Jalali and P. Mermelstein, "Effects of diversity, power control and bandwidth on the capacity of microcellular CDMA systems," *IEEE Journal on Selected Areas In Communications*, vol. 12, pp. 952–961, June 1994.

[10] W. Mohr and M. Kottkamp, "Downlink performance of IS-95 DS-CDMA under multipath propagation conditions," *Proc. IEEE ISSSTA*, vol. 3, pp. 1063–7, May 1996.

[11] P. R. Pawlowski, "Deorthogonalization of ODS-CDMA QPSK by AM/PM nonlinearity," *Wireless Personal Communication (Kluwer Academic Publishers)*, vol. 6, (no. 1-2), pp. 5–25, Jan. 1998.

[12] R. D. Gaudenzi and F. Giannetti, "DS-CDMA satellite diversity reception for personal satellite communications: satellite to mobile link performance analysis," *IEEE Trans. Veh. Tech.*, vol. 47, No. 2, pp. 658–72, May 1998.

[13] R. Stuetzle and A. Paulraj, "Modeling of forward link performance in IS-95 CDMA networks," *Proceedings of ISSSTA 1996*, vol. 3, pp. 1058–62, Sept. 1996.

[14] G. L. Stuber, *Principles of Mobile Communications*. Boston: Kluwer Academic Press, 1996.

[15] A. J. Weiss and B. Friedlander, "Synchronous DS-CDMA downlink with frequency selective fading," *IEEE Transactions on Signal Processing*, vol. 47, no. 1, pp. 158–67, Jan. 1999.

[16] N. Yuen and B. Friedlander, "Asymptotic performance analysis for signature waveform estimation in synchronous CDMA systems," *IEEE Trans. Signal Proc.*, vol. 46, no.6, pp. 1753–7, June 1998.

A New Multiuser Detector for Synchronous CDMA Systems in AWGN Channels[1]

Adrian Boariu and Rodger E. Ziemer
Electrical and Computer Engineering Department
University of Colorado at Colorado Springs
1420 Austin Bluffs Pkwy
Colorado Springs, CO 80907, USA
aboariu@eas.uccs.edu, rziemer@nsf.gov

Abstract – The decorrelating decision-feedback (DDF) multiuser detector based on Cholesky factorization has been proven to improve the performance of the users in the detection process. For relatively low crosscorrelation values between user signals this detector performs quite well. The detector described in this paper employs two triangular matrices (upper and lower) and soft output information to improve the data estimates over the DDF detector. Significant performance gains can be achieved over the DDF. Also, the users tend to have their bit error probabilities clustered. Thus, the performance of a certain user is less dependent on its position in the detection process than for the DDF.

1. Introduction

Several multiuser detectors have been introduced for synchronous CDMA systems. The decorrelating detector [1] employs the inverse of the crosscorrelation matrix of the spreading code in the detection of each user. The disadvantage of this detector is that it enhances the noise. The decorrelating decision-feedback (DDF) detector, proposed in [2], employs Cholesky factorization to determine decorrelating and feedback matrices for the multiuser detection process. The decorrelating matrix whitens the noise while the feedback matrix is used to detect the users successively. For best performance, the users are sorted according to their transmitted powers. Also, the strongest users are detected first followed by the weaker ones. Using this method, the strongest user gives the same performance as the decorrelating detector, while the weakest user gives the same performance as the single user lower bound if perfect multiuser interference cancellation is assumed. The overall performance of the DDF detector is better than the decorrelating detector.

[1] This research was supported by the Office of Naval Research under contract N00014-920-J0176 UP00004

An improved multiuser detector that uses two Cholesky matrices iteratively is presented in this paper. Soft output information is used in order to improve the data estimates.

2. System Description and the Cholesky-Iterative Detector

The system to be analyzed is shown in Figure 1. There are K users, each employing a specific sequence, c_k, for spreading with $c_k = [c_{1,k}\ c_{2,k} \dots c_{Q,k}]$ where Q is the spreading factor. Let $c = [c_1{}^T\ c_2{}^T \dots c_K{}^T]^T$ be the spreading code matrix of all users, where the superscript T denotes the transposition. The transmitted powers of the users are included in their spreading codes. The CDMA system is synchronous and the channel is assumed to be AWGN. The output of the matched filter bank at the moment kT_s, where T_s is the symbol duration, is given by

$$y_k = R\,x_k + z_k \tag{1}$$

where $R = c\,c^T$ is the crosscorrelation matrix of the spreading codes and z_k is colored noise with the covariance matrix $N_0\,R/2$.

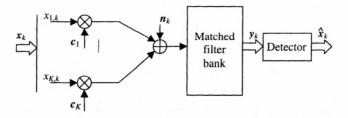

Figure 1. Synchronous CDMA system in AWGN channel

The Cholesky-iterative detector [3] introduced in this section employs two matrices obtained by Cholesky factorization. One is upper triangular and the other is lower triangular. For example, suppose we have

$$R = \frac{1}{7}\begin{bmatrix} 7 & -1 & 3 & 3 \\ -1 & 7 & -1 & 3 \\ 3 & -1 & 7 & -1 \\ 3 & 3 & -1 & 7 \end{bmatrix};\ f_1 = \begin{bmatrix} 1 & -0.14 & 0.43 & 0.43 \\ 0 & 0.99 & -0.82 & 0.49 \\ 0 & 0 & 0.9 & -0.31 \\ 0 & 0 & 0 & 0.68 \end{bmatrix};\ f_2 = \begin{bmatrix} 0.68 & 0 & 0 & 0 \\ -0.31 & 0.9 & 0 & 0 \\ 0.49 & -0.82 & 0.99 & 0 \\ 0.43 & 0.43 & -0.14 & 1 \end{bmatrix};$$

for the crosscorrelation matrix, R, and the upper and lower triangular matrices, f_1 and f_2, respectively, resulting from Cholesky factorization. If the f_1 matrix is used in the detection process, the first user will achieve its single-user lower bound for perfect cancellation of the interference, while the last user will be 3.35 dB worse than its single user lower bound. If the f_2 matrix is used instead, the situation is just the opposite; the last user will achieve its single user lower bound for perfect cancellation of the interference, while the last user will be 3.35 dB worse than its single-user lower bound.

The solution is to use both matrices simultaneously and to iterate the detection, alternately employing the f matrices. Of course, soft decisions are needed in order to take full advantage of the method. For a given user, two soft decisions are computed each time (for each f matrix), and the one that is most reliable is kept. Based on it, the data is estimated. The soft information that will be used in the detection process is the log-likelihood ratio, which for the ith user at the kT_s moment is

$$L_{i,k} = 4 f_{ii} \, y_{i,k}. \tag{2}$$

while the data is estimated based on

$$x_{ik} = \text{sign}(L_{i,k}). \tag{3}$$

The bit error probability (BEP), assuming perfect interference cancellation, is given by

$$P_i = Q\left(\max(f_{1,ii}, f_{2,ii}) \sqrt{\frac{2}{N_0}} \right), \quad i = 1, 2, ..., K. \tag{4}$$

The detector is a mixture of linear and nonlinear procedures. Soft outputs for each user can be provided if needed for subsequent signal processing.

3. Simulation Results

Simulations with four Gold codes of length seven, having the correlation matrix, R, given in the previous example, have been performed. Simulation results are presented in Figure 2 for the for DDF detector and in Figure 3 for Cholesky-iterative detector. Clearly, the Cholesky-iterative detector provides a gain over the DDF multiuser detector. The performance of a given user is less dependent on its position in the R matrix than for the DDF detector since the performances of the users are clustered.

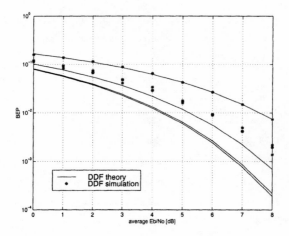

Figure 2. Performance of the DDF detector

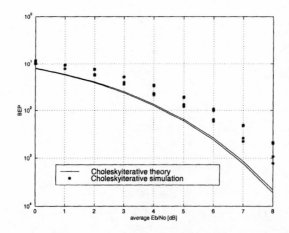

Figure 3. Performance of the Cholesky-iterative detector

4. Conclusion

The Cholesky-iterative detector was introduced in this paper. It employs two triangular matrices obtained through Cholesky factorization in order to improve the data estimates. Simulation results show that the Cholesky-iterative detector outperforms the well-known DDF detector. Also, the performance of a given user is less dependent on its position in the detection process than for alternative detector structures, i.e., decorrelating and DDF detectors.

REFERENCES

[1] R. Lupas and S. Verdu, "Linear multiuser detectors for synchronous CDMA channels," *IEEE Trans. on Info. Theory*, vol. 35, pp. 123-136, Jan. 1989

[2] A. Duel-Hallen, "Decorrelating decision-feedback multiuser detector for synchronous CDMA channel," *IEEE Trans. on Commun.*, vol. 41, pp. 285-290, Feb. 1993

[3] A. Boariu, *Multiuser detectors for synchronous CDMA communications systems in doubly spread channels*, Ph. D. dissertation, Univ. of Colorado at Colorado Springs, May 1999

Modeling Study to Determine the Realistic Constraints of the Wireless Land Mobile Radio Narrowband CAI Interface Specified in the TIA-102 Standard

Stephen E. Bartlett and Khalid M. Syed, Members, IEEE

Booz·Allen & Hamilton Inc.

8283 Greensboro Drive

McLean, VA 22102

Abstract

This paper recounts the outcome of a study of the channel performance and potential interoperability of the common air interface (CAI) proposed in the TIA-102 narrowband standard for public safety land mobile radios for voice and data communications. Acceptance of this standard has been based primarily on its capability to support 9600 bps data rates at both 12.5 kHz and narrower bandwidths, as well as compatibility between the standard's two types of transmitter specifications, C4FM and CQPSK, with a single type of CFDD receiver based on an FM discriminator. For this study, capabilities are measured using a simulated system of transmitters and receivers modeled in accordance with the TIA-102 standard. Wireless propagation parameters are introduced in the models to assess the standard's behavior over varying distances and vehicle velocities. The limits to channel capacity are determined for communications between each transmitter and receiver specified. Analysis is done using signal processing routines within SystemView. The study illustrates the standard's realistic technical feasibility, as determined by the model, that may be useful for vendors in the wireless community who may be considering the manufacture of equipment based on the TIA-102 standard.

1. Introduction

The TIA-102 standard is the proposed open standard for the next generation of digital narrowband two-way land mobile radio (LMR) systems. The TIA-102 standard, which is also known as the APCO Project 25 standard, has been in development since 1989, was accepted by the TIA standards committee, and has a newly released edition dated June 1998 [1]. Adoption of the TIA-102 standard has been predicated on both a backward compatibility to legacy analog systems, and an ability to foster interoperability in the public safety land mobile radio user's community (see page 30 of TSB102 *System and Standard*

Definitions in [1]). Additionally, the standard defines a common air interface for narrowband digital radio, which is backward compatible between the different proposed technical implementations of the standard itself (see page 12 of TSB102 *System and Standard Definitions* in [1]). Development of phase one of this standard is under way for 12.5 kHz bandwidths. The second phase of the TIA-102 standard has not yet been defined per se, but it has been suggested that it be defined for a 6.25 kHz bandwidth. The current standard does not specifically differentiate between the 12.5 kHz channel and a future narrowband 6.25 kHz channel requirement (it only specifies the capability of "migrating to narrower channel spacing..."). The feasibility for full interoperability and gross channel data capacity of 9600 bps of the two transmitter modulation schemes, Compatible 4-level Frequency Modulation (C4FM) and Compatible Quadrature Phase Shift Keying (CQPSK), and the single TIA-102 Compatible Frequency Discriminator Detection (CFDD) receiver design is examined in this paper. The C4FM modulation scheme for 12.5 kHz channel spacing is the most common economical design thus far, most likely due to the constant envelope modulation characteristic which, while being spectrally efficient, also makes the use of nonlinear final amplifiers feasible. For a narrower channel phase, use of the CQPSK modulation scheme or an equivalent technology, especially for a 6.25 kHz channel spacing [2,3], will be necessary. Section 2 describes the TIA-102 simulation models and the assumptions made to develop them, within the authors' interpretation, in accordance with the standard, which was our primary reference source. The model was measured in accordance with the TIA-102 T102BAAB *Common Air Interface Conformance Tests,* where appropriate. Section 3 discusses the simulation measurement results for the transmitter and receiver designs with various data rates as a function of signal to noise (Eb/No) in an AWGN channel, sampling time jitter, and potential Doppler effects for different frequency bands with a vehicle velocity of 120 km/hr.

2. Description of Model

The TIA-102 12.5 kHz narrowband digital two-way radio model was simulated using SystemView software by ELANIX. The simulated model consisted of a C4FM and CQPSK transmitter and an FM discriminator receiver, each modeled in accordance with the TIA-102 standard. An additional noncoherent quadrature receiver was also modeled to fully assess the CQPSK transmitter. The transmitters and receivers were linked via a variable AWGN wireless channel simulator for selectable Eb/No parameters. Data throughput was evaluated by measuring the BER performance under different conditions. The simulation model operated at a sampling rate of 96 kHz, with the simulated channel operating at 24 kHz (1/4 the sampling rate) to ensure that no aliasing would occur during simulation runs.

Transmitters - The transmitters were modeled in accordance with the ANSI/TIA/EIA102.BAAA *Project 25 FDMA Common Air Interface* section of the standard, which defines the general characteristics of the two modulators in the context of a QPSK-C family, which is described as a blend of 4-level FSK and a form of π/4Differential Quadrature Phase Shift Keying (π/4DQPSK). The standard defines the transmitter that modulates the phase but keeps the carrier amplitude constant as a C4FM, and the transmitter that modulates the phase and amplitude of the carrier simultaneously as a CQPSK. [1] This CQPSK modulation scheme in unclear considering MPSK modulators are constant amplitude, unless filtered. The CQPSK and C4FM transmitter models designed for the simulations are shown in figures 1 and 2 respectively.

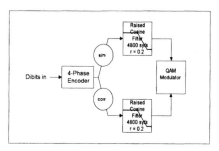

Figure 1: CQPSK transmitter simulation model

Figure 2: C4FM transmitter simulation model

Symbols - The symbol rate of the modulators is defined to be 4800 symbols per second yielding a symbol period of 208.33 microseconds, each symbol conveying 2 bits of information with the following mapping as specified in the standard [1]:

Table 1: TIA-102 Bit-to-Symbol Mapping

Information Bits	Symbol	CQPSK Phase change	C4FM Deviation
01	+3	+135 degrees	+1.80 KHz
00	+1	+45 degrees	+0.60 KHz
10	-1	-45 degrees	-0.60 KHz
11	-3	-135 degrees	-1.80 KHz

The encoders shown in figures 1 and 2 conform to this symbol convention.

Nyquist Raised Cosine Filter – To constrain the channel bandwidth and reduce the Intersymbol Interference (ISI) that generally occurs in narrowband channels, pulse shaping is necessary and is

152

addressed through the specification of a raised cosine filter. The standard specifies the following characteristic equation for the raised cosine symbol filter:

f = frequency in Hertz

Passband |f|<2880 Hz
|H(f)| = magnitude response of the Nyquist Raised Cosine Filter
|H(f)| = 1   for |f| < 1920 Hz
|H(f)| = 1/2[1+cos(2πf/1920)]   for 1920 Hz <|f|<2880 Hz
|H(f)| = 0   for |f| > 2880 Hz.   1

where the rolloff factor α was derived from the filter boundary condition:

$$(1+\alpha)/2Ts = 2880$$ 2
$$\alpha = 0.2$$

In this model, the raised cosine filter uses a rolloff factor of α = 0.2 with the specified passband of 2880 Hz. Figure 3 depicts the response curve.

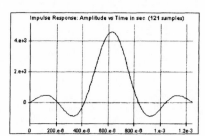

Figure 3: Raised Cosine Impulse Response

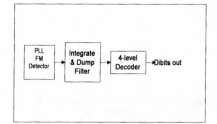

Figure 4: TIA-102 FM Detection Receiver

Receiver - The TIA-102 standard defines a Compatible Frequency Discriminator Detection (CFDD) receiver design capable of receiving a signal either from the C4FM modulator or the CQPSK modulator[1]. This specification, along with the requirement to be backward compatible to legacy analog FM systems, implied to the authors that the noncoherent FM discriminator receiver design, described in the standard, is all that is necessary for both types of TIA-102 CAI modulators. FM discriminators are not necessarily a good match for amplitude and phase modulated transmitters as the CQPSK is described in TIA-102, but have been shown to work suitably for differentially encoded π/4 DQPSK, and Generally Tamed Frequency Modulated (GTFM) transmitters (the binary equivalent of the TIA-102 specified C4FM) [2, 3, 4].

The receiver specified in the ANSI/TIA/EIA102.BAAA *Project 25 FDMA Common Air Interface* closely resembles the noncoherent FM discriminator design studied in [3]. The simulation model used for this study is also a similar design to that in [3] and is shown in figure 4.

Previous studies in narrowband digital modulation and detection techniques have shown that results for noncoherent receivers, although the suboptimal design, do work well in the fast fading environment that a mobile radio suffers. Noncoherent receivers can circumvent the fast carrier recovery problems suffered by coherent receiver designs, however, such designs do result in poorer BER performance than coherent designs.

Channel Simulation - An AWGN channel was used to simulate signal to noise parameters, along with fading and doppler effects, to ascertain the speed and distance effects from the transmitter to the mobile. Specific doppler effects are dependent on the RF frequency and the velocity of the mobile receiver, as given by the relation:

$$fd = fc \frac{v}{c} \cos \theta \qquad\qquad 3$$

where (v) is the velocity relative to the angle from the transmitting antenna radial given by $\cos \theta$, fc is the radio carrier frequency, and c is the velocity of light in free space. In this study, a vehicle velocity of 120 km/hr was used to model Doppler phase changes at 16, 50, and 90 Hz representing different frequency bands as shown in table 2:

Table 2: Doppler shift vs. frequency for 120 km/hr vehicle velocity

Radio Frequency	*Velocity (km/hr)*	*Maximum Doppler Shift (Hz)*
150 MHz	120 km/hr	16
450 MHz	120 km/hr	50
800 MHz	120 km/hr	90

Special Design Constraints - Voice signal encoding with the Improved Multiband Excitation (IMBE) vocoder, as specified in TIA-102, is not used in this simulation. However, indirect assessments are made by measuring the channel with a bit rate of 7200 bps as specified for the vocoder output in ANSI/TIA/EIA102.BABA *Project 25 Vocoder Description.*

The TIA-102 C4FM transmitter design specifies the need for an additional pulse shaping filter following the Nyquist raised cosine filter with the characteristic response of $|P(f)| = (\pi f/4800)/\sin(\pi f/4800)$, for $|f| <$

2880 Hz, with a flat group delay over the passband. This shaping filter is not included in the model's design, because the Nyquist raised cosine shaping filter is fed with impulses instead of a non-return to zero (NRZ) waveform. Consequently, the removal of the additional $\sin(\pi f/4800)/(\pi f/4800)$ signal, which an NRZ waveform would create, is not necessary.

The TIA-102 symbol mapping illustrated in Table 1 is a nonorthogonal symbol convention for the C4FM modulator. A test for noncoherent M-ary orthogonal signals involves checking that the tone spacing is chosen to be some integer multiple of the symbol time Ts $(Ts=1/4800)$. This ensures that the symbol frequencies are uncorrelated over the symbol interval Ts[7]. To test for this, the product of the TIA-102 C4FM tone spacing and symbol time Ts should yield an integer number, $dfm*Ts = integer$ [8]. The symbol tone spacings are: 1200Hz (+1,-1 or +1,+3), 2400Hz (+1,-3), and 3600 Hz (+3,-3), which result in the respective $dfm*Ts$ products: ¼, ½, ¾. Thus, these fractional products (non-integers) indicate that the C4FM symbols are not an orthogonal set, as defined for a noncoherent M-ary signal. As will be seen, this contributes to a poor BER performance characteristic for this modulation scheme.

The C4FM signal is successfully received and decoded by the CFDD receiver as modeled in this study. However, modeling this CFDD receiver to also detect the CQPSK modulation, as described in the TIA-102 standards document, was not successfully accomplished. It was decided that the CFDD receiver design, as described by the standard, is silent on the specifics of this design feature. As an alternative, the system is modeled using both the CFDD FM discriminator receiver, for use with the C4FM transmitter, and a noncoherent quadrature receiver built specifically to match the CQPSK transmitter. The simulation and testing were then done on each system as a complete transmitter-receiver pair.

Finally, a bit convention that determines whether the most significant bit (MSB) or least significant bit (LSB) of each dibit pair should be converted first is not addressed in the standard. For consistency in the models used for this study, the authors chose an "LSB first" convention. Note: if two manufacturers differ in this preliminary design choice, their equipment will be incompatible.

3. Simulation Results

Bit Error Rate Performance - The data rates representing the gross channel throughput of 9600 bps and the IMBE throughput of 7200 bps, are passed through the system model to assess the BER performance versus the Signal to noise ratio (Eb/No) for each modulation scheme.

The C4FM transmitter is paired with the FM discriminator receiver, and the CQPSK transmitter is paired with a quadrature receiver for BER performance comparisons using an RF channel simulation, as illustrated in figure 5, which shows the model used for the CQPSK system tests.   In this channel simulation, the AWGN channel noise is incrementally reduced to simulate increasing signal to noise ratio (Eb/No) for measuring the BER vs. Eb/No performance curves shown in figure 6. The TIA-102 standard document TSB102-CAAB *Digital C4FM/CQPSK Transceiver Performance Recommendations* specifies the minimum BER, at the reference sensitivity, to be 5%, or an error probability not to exceed 5 x 10 $^{-2}$. As can be seen from the curves in figure 6, the C4FM system does not meet this specification until the Eb/No ratios are greater than 14 dB. This is a poor BER performance for this C4FM/CFDD transceiver pair, most likely due to the fact that the signaling states, as described earlier for the C4FM modulation, are non-orthogonal. As a consequence, the symbol frequencies are too close together in symbol time to be uncorrelated and overlapping occurs, causing symbol detection errors in the presence of channel noise. Another contributing factor is additional inter-symbol-interference (ISI) introduced by the integrate and dump filter in the CFDD receiver.

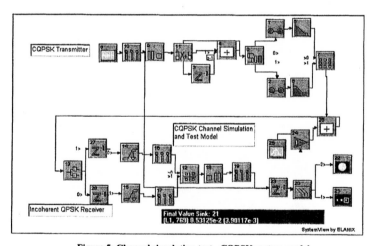

Figure 5: Channel simulation test - CQPSK system model

156

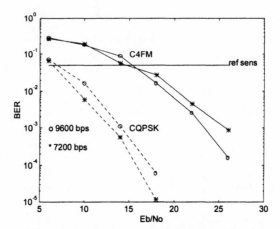

Figure 6: BER performance of C4FM and CQPSK models

Referring again to figure 6, it can be seen that the BER performance for the CQPSK system is clearly superior to that of the C4FM. Comparing the reference sensitivity points for each modulation scheme, the CQPSK system shows a 7 dB BER performance improvement over the constant envelope C4FM system. The CQPSK transceiver pair benefits from the biorthogonal signaling used in the CQPSK differential modulation scheme, which are easily detected in a noisy channel due to noncorrelated symbols with no overlapping.

Timing Errors – Timing error jitter testing on the C4FM and CQPSK system models is run to study the sensitivities to timing jitter BER performance. The results of these tests are graphed in figure 7. The tests are run by setting each system at the reference sensitivity level of 5% BER. Next, the sampling time jitter at the receiver is stepped by a time increment dt = 0.01τ, or 1/10 bit time (1/τ = 9600). Jitter sensitivity is compared by defining a threshold where the jitter time value (dt) causes the BER to exceed the 5% threshold. As can be observed in figure 7, the C4FM modulation, with a timing stability of 8% τ, is more jitter tolerant than CQPSK, which shows a timing stability of 5% τ. This result agrees with the findings in [2]. In general, the jitter sensitivities are likely due to the low alpha rolloff value of 0.2 in the raised cosine filters. Although this rolloff value is necessary for the reduction of ISI and the narrow channel bandwidth, it may be limiting the system tolerance to jitter.

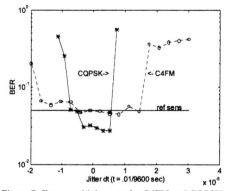

Figure 7: Jitter sensitivity tests for C4FM and CQPSK models

C4FM Doppler Tests - The Doppler frequency shift is modeled for the C4FM system to determine to what extent rapid vehicle motion could further degrade the BER performance. The results are graphed in figure 8. The test simulates a vehicle driving towards a transmission tower, at a constant radial angle, and a vehicle velocity of 120 km/hr. This results in an increase or blue-shift in the carrier frequency, which, if not well tracked by the FM discriminator receiver, will create a DC offset in the detected data stream and further degrade the BER performance of the channel. To illustrate this effect over a span of frequencies, the frequency bands of 150 MHz, 450 MHz, and 800 MHz were simulated. Full Doppler spreading from multipath signals is not modeled in this test. As can be seen in figure 8, the 800 MHz band suffers the most BER performance degradation from a Doppler shift at the receiver. As a comparison, the 150 MHz frequency appears to experience the least BER reduction at the 120 km/hr velocity, with a difference between the 150 MHz and 800 MHz bands of 3-4 dB at a BER of 1×10^{-3}. One consequence of this effect could occur at the fringes of a C4FM system where rapid vehicle motion, resulting in higher data error, could cause dropouts or numerous retries from the sudden change in BER performance.

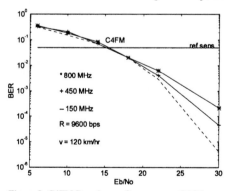

Figure 8: C4FM Doppler performance vs. CAI frequency

158

Spectral Signatures – The unfiltered power spectral responses for the transmitter models are shown in figure 9, with the TIA-102 12.5 kHz spectral mask outlined in red. This study did not investigate the full impact of adjacent channel interference (ACI) between the two modulation schemes. The significance of potential spectrum overlaps will become more apparent for two TIA-102 transmitters operating on adjacent channels, but each employing a different modulation. Special considerations for this possibility are needed for future TIA-102 narrower band systems (bandwidths <12.5 kHz). The C4FM will most likely be operating at 12.5 kHz bandwidth, with a transmitter that does not employ linear power amplification. Conversely, the narrower band systems will be operating with CQPSK, by necessity, and will need to employ linear power amplification to suppress the harmonic spikes, which can be seen in the CQPSK spectrum in figure 9. There have been studies of the potential for ACI between two spectral models, similar to the TIA-102, identified as GTFM and DQPSK in[2]. A study of adjacent channel interference between these two different modulation envelopes, using an 8kbps gross data rate, showed that for adjacent channels of constant and variable envelope modulations with 12.5kHz and 6.25kHz spacing, the DQPSK system offered superior ACI protection against the GTFM, but, conversely, the GTFM suffered significantly reduced ACI protection from the DQPSK modulation. The TIA-102 specifies a higher gross data rate of 9.6 kbps, and experiences even greater ACI than the modulation modeled in [2].

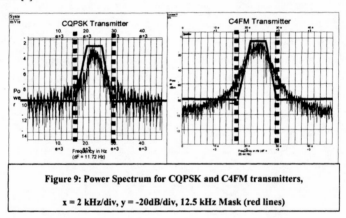

Figure 9: Power Spectrum for CQPSK and C4FM transmitters,

x = 2 kHz/div, y = -20dB/div, 12.5 kHz Mask (red lines)

4. Summary and Conclusions

The standard specifies two modulation schemes for implementing two narrowband 12.5 kHz spaced channels from the splitting of one 25 kHz channel. Both schemes employ a type of differential, four level modulation; a constant envelope frequency modulation (C4FM); and a phase and amplitude modulation (CQPSK). It has been found that the simple model based on the TIA-102 standard produces a nonorthogonal C4FM transmitter system with poor BER performance characteristics. The C4FM operating in the 800 MHz spectrum suffers a worse BER performance degradation at high speeds, from Doppler effects, than the 150 MHz band spectrum. Both systems are sensitive to timing errors, but the C4FM system experiences less sensitivity to timing errors than the CQPSK. Both transceiver systems show a slightly different BER performance between the IMBE required 7200 bps data rate over the 9600 bps gross channel data rate. C4FM also shows a reduced BER performance at 7200 bps, possibly due to the nonorthogonal symbols.

Although the TIA-102 standard specifies a common CFDD receiver capable of receiving either modulation scheme, it was found in this study that this receiver concept cannot be easily employed. This problem has been reported in earlier reports where manufacturers involved in implementing this standard have stated, "in the compatible receiver, AM/PM conversion in the RF and IF stages and level changes caused by differences in phase deviation between the C4FM and CQPSK signals, are manufacturing issues that must be resolved." [6] If given a choice between the two modulators, most manufacturers will defer to the more economical design which is the C4FM transmitter with a nonlinear power amplifier and the cost effective FM discriminator receiver, which, may not be designed to effectively receive the CQPSK signal. As a consequence, any manufacturer employing the more complex CQPSK design in the future may be inoperable with these C4FM and FM discriminator transceiver systems. Therefore, should the migration to a narrower band spectrum occur, and the CQPSK modulator be adopted for these narrower band systems of the future, a radio system using the 12.5 kHz C4FM/CFDD may not be compatible with these narrower band systems and will most likely have to be summarily replaced. This is much the same as today's migration to the 12.5 kHz narrowband digital systems from the 25 kHz legacy analog systems.

Future Study – Future study is needed to define the optimal design for the CFDD receiver that effectively receives both C4FM and CQPSK signals equally well. This would help to define a more open receiver design for the standard. More extensive study is necessary to determine the full effects of the TIA-102 channel in an environment that includes varying multipath and full Doppler spreading effects with rapidly changing mobile velocities. Of special interest is the effectiveness of this standard in crowded spectrum where a variety of TIA-102 implementations have been deployed. Modeling the IMBE vocoder model with TIA-102 DES-OFB digitally encrypted voice would also be of value. Further modeling of the TIA-102 systems and their various features in narrower bandwidth channels is a priority for future migration strategies. These studies would reveal the TIA-102 standard's capabilities in rigorous realistic environments that the public safety users in the field face regularly.

5. Acknowledgements

The authors wish to thank Maury Schiff, chief scientist at ELANIX, for his generous help and time saving suggestions during the development of the models used in this study. The ELANIX support during our learning process was welcome and needed in finding the many valuable functions and features SystemView provides.

VI References

1.   Project 25, The TIA-Published 102-Series Documents, June 1998.

2.   S. Pizzi, H. Gibbons, M. Morrissette, C. Wright, "Comparison of Generalized Tamed-FM and Differentially-Coherent QPSK for Land Mobile Radio, *IEEE Transactions on Communications,* April 1994.

3.   K.S. Chung, "Generalized Tamed Frequency Modulation and Its Application for Mobile Radio Communications", *IEEE Transactions on Vehicular Technology*, August 1984.

4.   T. S. Rappaport, *Wireless Communications Principles and Practice*, New Jersey, Prentice Hall, 1996.

5.   W. Stuble, F. McGrath, K. Harrington, P. Nagle, "Understanding Linearity in Wireless Communications Amplifiers", *IEEE Journal of Solid State Circuits,* September 1997.

6.   Chase, Alan W., "A Glimpse at the New Technology Under Study by APCO Project 25, *The Journal of Public Safety Communications*, August 1991.

7.   B. E. Keiser, *Broadband Coding, Modulation, and Transmission Engineering.* New Jersey: Prentice Hall, 1989.

8.   Stephen G. Wilson, *Digital Modulation and Coding. .* New Jersey: Prentice Hall, 1996.

Over-The-Air Subscriber Device Management
Using CDMA Data and WAP

Nadine L. Marran

Bell Atlantic Mobile
180 Washington Valley Road
Bedminster, NJ 07921
MarraNa@bam.com

Abstract

This paper describes an architecture for Over-The-Air Service Provisioning (OTASP) and Over-The-Air Parameter Administration (OTAPA), which is an alternative to IS-683A and IS-725-A. The architecture uses standard CDMA data and the Wireless Application Protocol (WAP), to communicate programming information between the CDMA digital handset and OTA application server. The handset WAP browser will serve as a friendly user interface, during the process. Wireless service providers should consider OTA applications as way to maximize the benefit of data services deployments. Additional applications planned for this environment are Over-The-Air Software Download (OTASD) and Over-The-Air Mobile Diagnostics (OTAMD). These applications will also be discussed.

The motivation to develop an alternative to IS-725-A and IS-683-A, is to take advantage of trends in handset technology, rather than further complicate handset requirements. The solution depends heavily on existing standards, which reduces the cost of deployment and increases the return on technology investment. This architecture is flexible enough to meet many long-term needs of wireless service providers. A subgroup of the CDG Steering Committee is currently drafting the specification document related to this OTA proposal.

Introduction

The booming wireless industry has created a very competitive market for wireless service providers. To meet that challenge, wireless carriers are working to expand their service offerings and improve customer satisfaction, while reducing operating costs and improving profitability. OTA Subscriber Device Management is one way to address those requirements. This technology includes a number of applications designed to automate routine processes and improve operating efficiency for wireless networks.

The OTA Concept

OTA is an acronym for Over-The-Air. It describes a type of application, which uses the wireless network for connectivity. OTA Subscriber Device Management describes a family of applications, used by carriers to manage digital subscriber units via the wireless network. These applications provide a wireless service provider with the ability to program, upgrade or troubleshoot a remote subscriber. Some examples are listed below:

Application	Description
OTASP (Over The Air Service Provisioning)	Facilitates the initial programming of subscriber device for a new customer activation or hardware upgrade. This function replaces manual programming of handsets at the point of sale. The session is initiated at the subscriber unit.
OTAPA (Over The Air Parameter Administration)	Facilitates the programming updates for previously provisioned subscriber devices. This function replaces manual re-programming of handsets associated with MIN changes, area code splits and preferred roaming list updates. The session can be initiated at the subscriber unit or the at the OTA application server.
OTAMD (Over The Air Mobile Diagnostics)	Facilitates the query or initiation of preprogrammed functions in the subscriber device to determine its current operational status or that of the system connection. This function provides the remote troubleshooting capabilities. The session is initiated at the at the OTA application server.

OTASD (Over The Air Software Downloads)	Facilitates the download of executable code to the subscriber device. Types of code that may be downloaded include the operating software of that device, or an individual application. The session can be initiated at the subscriber or at the OTA application server.
OTA E-Care (Over The Air Electronic Customer Care)	In addition to toll-free hotlines, most organizations also offer customer service and support via the WWW. This application is a text-only version of the customer care website, specially formatted for the subscriber screen display. The session is initiated at the subscriber unit.

While these basic functions are common for managing elements of wired networks, we are just starting to explore OTA versions. Fortunately, the OTA concept is quickly gaining the support necessary for its development. The wireless industry is beginning to shift toward a software-based approach for network design and management. To demonstrate the significance of this shift, the traditional model for wireless subscriber management will be described.

Subscriber Management Today

There are minimum requirements for supporting a mobile subscriber device. Assuming that a customer is equipped with functioning subscriber hardware and software, which is compatible with the service provider network, the conditions required for service are that:

- The mobile device is programmed with the appropriate network parameters (i.e. - Mobile Telephone Number, Home System ID, etc.)
- The service provider network is provisioned with the appropriate subscriber account information

Mobile device programming is typically a manual process. Some drawbacks to manual programming are the resources required and the opportunity for error. Because of limited resources, especially during the peak retail season, this process is limited to minimal programming of required parameters. Additional parameters, such as Preferred Roaming Lists (PRLs), may be pre-programmed into the device by the manufacturer. Since many parameters are network specific, custom preprogramming requires a special arrangement between the service provider and manufacturer.

After the initial programming of the subscriber, parameter updates may be required. If the customer requests a new Mobile Telephone Number or area codes are reassigned in that service area, the subscriber device must be reprogrammed. There are other occasions when parameter updates are desired to enhance performance or network efficiency. An example is the Preferred Roaming List, which may change as the result of a merger between wireless service providers.

There are several other trends in the wireless industry, which are driving the need for OTA Subscriber Device Management. The first is the rapid growth of the wireless subscriber base. As the number of subscribers continues to grow, the more subscriber devices there are in the field.

U.S. Wireless Subscribership: December 1985 - December 1998

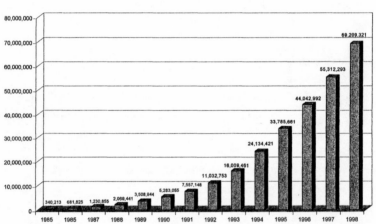

Source: CTIA

Mobile provisioning has also become more complex in recent years, as nation-wide service plans, short messaging and even Internet access become more common mobile service offerings. There are also a growing variety of mobile subscriber devices, from smart phones and wireless PDAs to low-end modules used for telemetry. Finally, increased competition is driving carriers to both reduce operating costs and improve customer satisfaction. All of these factors are creating a great opportunity for the automation of handset management processes.

Limitations of Current Standards

There are existing standards defined for OTASP and OTAPA. IS-725-A defines the message structures and commands for OTASP and OTAPA capabilities. IS-683 is the air interface standard for OTASP and OTAPA in CDMA systems. IS-683A OTA operations occur between a mobile device and an Over-The-Air Function (OTAF) using IS-95A traffic channel data burst messages. This technology has been available for two years, but the implementation to date has been limited.

These standards for OTA are centered on voice service programming and are not flexible enough to accommodate the full range of carrier needs. The standards to do not address programming of data parameters at this time. Each new feature or enhancement will require development for the network infrastructure, the mobile device and OTAF. Use of the messaging channel as transport does provide the ability to download parameters during a voice call, which may be considered as the best feature. Although the simultaneous voice session may be desirable at times, it is an added complication for carriers looking for to fully automate the OTASP process. That feature also limits data throughput during the OTA session. In order to support future applications such as OTA Software Download (OTASD) more bandwidth is required. The long-term drawbacks of the IS-725-A and IS-683 standards have led CDMA service providers to seek alternative solutions.

The Solution

This proposal for OTA Subscriber Device Management requires that the CDMA carrier have an IS-99 data services-capable network. IS-99 is the original CDMA data standard published in 1995. It defines asynchronous data (and digital Group 3 fax) over a circuit switched connection. Asynchronous data basically allows a cell phone to provide the exact same service as a dial up telephone modem. In the future, these OTA applications could be migrated to packet-based transports as well.

Another important component of the OTA solution is the Wireless Application Protocol (WAP). WAP is an open wireless protocol specification based on Internet standards such as XML and IP. The WAP programming model is similar to the WWW programming model. This provides several benefits to the application developer community, including a familiar programming model, a proven architecture, and the ability to leverage existing tools (e.g., Web servers, XML tools, etc.). Optimizations and extensions have been made in order to match the characteristics of the wireless environment. Wherever possible, existing standards have been adopted or have been used as the starting point for the WAP technology. A micro browser in the mobile device provides the user interface and is analogous to a standard web browser.

The solution architecture is depicted in the diagram below. Each component is discussed in greater detail in the following sections.

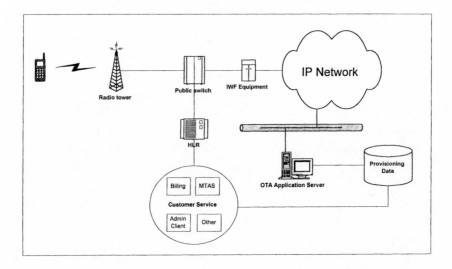

In this architecture, the mobile device originates a data call, which is routed to the Inter-Working Function (IWF) via the switch. A PPP connection is established between the handset and the IWF. The IWF connects the mobile to the WAP gateway and an IP connection is established between the mobile and the gateway. The mobile device requests a specific URL and the gateway satisfies the request by connecting to the Provisioning Server and requesting the URL. The requested resource is sent back to the micro-browser in the mobile subscriber device, where it is processed. The provisioning data is extracted and written to the proper memory stores.

The cell site, mobile switching center and IWF provide the conduit through which the MS, WAP gateway and provisioning server can communicate IP-OTASP and IP-OTAPA requests and responses. These last three components form the framework for IP-based OTA application environment.

The subscriber device is a client to the WAP gateway and as such, it must have a WAP browser and the WAP protocol stack. The fact that this connection is established via the CDMA Data link and IWF requires that the MS also support RLP, PPP and IP. The WAP browser must support the Wireless Markup Language (WML) and must be capable of updating NAM parameters, roaming lists and A-Key parameters. This requires standard specifications for embedding provisioning data in WML content and updating the MS's memory.

The WAP gateway translates requests from the WAP protocol stack to the WWW protocol stack. Additionally, the gateway is responsible for controlling access to a given carrier's Intranet. It will ensure that only those customers who request a particular service are authorized for it.

The OTA Application Server resides on an IP network within the controlled confines of the carrier. The OTA Application Server must have an interface to the carrier's Mobile Terminal Authorizing System (MTAS) and billing system. This interface serves to synchronize the OTA Application Server to the information in the MTAS and account records. The specific requirements of this interface are dependent on the capabilities and interfaces of the carrier's customer care center system(s). The OTA Application Server must be capable of receiving dynamic updates from the MTAS and have the provisioning information immediately available for downloading into the chosen mobile station or groups of mobile stations.

Deployment Issues

Although this solution depends heavily on existing standards such as IS-99 and WAP, standardization is a required for the proposed OTA data structures and interface specifications. The process of standardization could take some time, which may delay the availability of OTA products. Handset development and certification for OTA support is a critical milestone, because it determines when and how quickly a carrier can populate the subscriber base. As more devices support this technology, carriers will be able to take advantage of OTA applications.

Interoperability is another challenge for OTA deployments. In order to support OTA services while roaming, carriers must deploy data services and establish interoperability requirements. IS-99 data technology is relatively new. Service providers must work together to define inter-network routing policies for data services.

Finally, IS-99 data does not easily accommodate Class of Service (COS) routing to establish a Virtual Private Network (VPN). OTA applications require special security and billing support. The ability to create a VPN over the data network would allow carriers to restrict OTA traffic and provide special call handling. IS-707-A packet data will most likely solve this issue.

The Road Ahead

The immediate goal of OTA management is to automate the existing manual processes for handset programming with OTASP and OTAPA. The purpose for this is to reduce operating costs, increase programming accuracy and improve network performance. Longer term, this same platform will be used to support Over-The-Air Mobile Diagnostics (OTAMD) and Over-The-Air Software Downloads (OTASD).

OTAMD allows the carrier technical support team to troubleshoot a remote subscriber device. This application reduces the time and cost associated with handset malfunctions and trouble resolution. It also improves customer satisfaction, as a quick and convenient way to address handset problems. Support for OTAMD requires software development in the subscriber device. Many of the functions are already available at the device, but require the interface for a remote IP-based server via the data connection. This application will be more powerful in the packet data environment when a data session may be set up during a voice call, allowing the technician to talk with the customer through trouble resolution.

OTASD will provide the ability to download executable code to mobile units via the wireless network. This application can be used to update handset software or add applications such as games, calendars, phone lists, etc. Customers can obtain the latest software without going to an authorized retailer. OTASD can be employed to facilitate a wide range of service offerings. However, it requires that digital subscriber devices have enough memory to support two copies of operating software and any additional software applications. Therefore, OTASP handset support may be delayed until the cost to extend handset memory is reduced.

By automating specific functions with these OTA applications, service providers have an excellent opportunity to integrate them with expert systems. A rule-based system can be put in place to synchronize provisioning of the network with programming the mobile device. This system will automatically cross check parameters to avoid a mismatch. Another example is "smart" roaming. "Smart" roaming uses an advanced set of rules to trigger the download of a preferred roaming list (PRL) to a mobile device. It is desirable to have a PRL that provides as much information as possible to ensure that a mobile selects the best network while roaming. However, a lengthy PRL may take up more memory than available in the handset or cause a delay when a mobile searches the PRL for a roaming partner. An expert or rule-based system will allow the PRL to adapt to the roaming needs of an individual customer. The system does this by cross-referencing the roaming profile of that customer with recent roaming history. The network will "learn" the roaming habits of the user, and build a PRL to match. Roaming into a new city may trigger an automatic OTAPA download for a PRL update.

An important benefit of this IP-based OTA solution is that the architecture can be used to offer other applications to wireless subscribers. This architecture provides the basic infrastructure, access, routing and security required by many wireless applications. Some services, such as e-mail, provide revenue opportunities while other services, such as E-Care, reduce operating expenses and improve customer satisfaction.

Conclusion

Wireless service providers and their customers can benefit by the deployment of OTA management services. In order to support these applications, wireless data services and WAP must become standard implementations on digital networks. In addition, the existing OTA standards must be updated to include IP-based message formats. New standards development is also required for OTAMD and OTASD. The growing support for OTA applications indicates that wireless industry is shifting toward a more software-based approach for network design and management. This step is very promising, and will provide the foundation for future advancements in wireless technology.

Acknowledgments

The content of this paper is largely derived from the development efforts of Bell Atlantic Mobile Network Engineering and the CDMA Development Group (CDG) subgroup on IP-Based Over The Air (IOTA) Handset Configuration Management.

Related Websites

http://www.bam.com
http://www.cdg.org/
http://www.tiaonline.org
http://www.wapforum.org/
http://www.wow-com.com

References

IS-99, 9.6 Kbps Data Services, Async Data + G3 Fax. Published in July 1995.

IS-637, Short Message Service (SMS), Rate Set 1. Published in December 1995.

IS-657, 9.6 Kbps Packet (I-net + CDPD). Published in July 1996.

IS-658, InterWorking Function. Published in July 1996.

IS-683, Over-the-Air Service Provisioning of Mobile Stations in Wideband Spread Spectrum Cellular Systems. Published in February 1997.

IS-683-A, Over-the-Air Parameter Administration of Mobile Systems in Spread Spectrum Systems. Published in September 1998.

IS-707, (14.4 kbps) Data Service Options for Wideband Spread Spectrum Systems. Published in February 1998.

IS-725-A, IS-41-C Enhancements for OTASP Over-the-Air-Service-Provisioning. Published in June 1997.

Hyperactive Chipmunk Radio

G.H. McGibney and S.T. Nichols

TRLabs

Suite 280, 3553 - 31 Street N.W.
Calgary, Alberta, Canada T2L 2K7

Email: grantm@cal.trlabs.ca

The hyperactive chipmunk radio modulates voice signals so that the radio waves behave the same in the radio medium as sound waves normally do in the acoustic medium. This is accomplished by segmenting the voice signal and compressing the segments in time before transmitting them through the radio channel. If the compression factor is correct, the distortion characteristics of the radio channel will match those normally encountered in the acoustic channel and sound natural to the ear. The radio signal will then inherit many of the good properties of acoustic voice signals including resistance to flat fading and tolerance of frequency selective fading.

1. Introduction

Human verbal communication has evolved into a very effective method of carrying information via sound waves through the multipath acoustic medium. Now as radio systems are used to extend the range of verbal communications, technological solutions are used to deal with the problems of the multipath radio medium, including flat fading and intersymbol interference. The hyperactive chipmunk radio avoids the need for technological solutions to these problems by using the natural human ability to deal with the acoustic medium to cope with the radio medium as well.

The design of this radio system is based on two observations: that radio channels and acoustic channels have similar characteristics; and that the human voice makes a very effective spread spectrum signal in the acoustic domain. Based on these observations, the hyperactive chipmunk method is described which allows the voice to be modulated onto a radio carrier in a way that translates the spread spectrum properties from the acoustic medium to the radio medium. The advantages and disadvantages of such a system are discussed.

2. Properties of Radio and Acoustic Channels

Both the radio and acoustic media produce channels with multipath characteristics. Figure 1 demonstrates the similarity between the measured frequency responses of the acoustic and radio channels. The channels are created when sound and radio waves propagate not just in a direct path from the transmitter to the receiver, but also by reflections off objects in the environment. Reflected signals must travel a further distance than direct signals, therefore they arrive at the receiver later in time. The composite of all the signals from the different paths, each with a different amplitude and delay, make up the multipath channel. For a detailed discussion of the properties of multipath channels see Proakis [1].

The behavior of a signal in a multipath channel depends on whether it is wideband or narrowband. Narrowband signals experience little or no distortion as they pass through multipath channels (other than additive white noise), however flat fading in the narrowband channel causes the received power level to fluctuate drastically and the signal to be lost entirely at times. In contrast, the overall power level of wideband signals is relatively stable. Across the

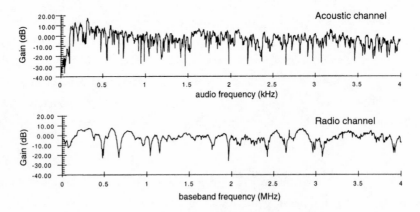

Figure 1. Measured frequency responses of acoustic and radio channels.

spectrum of the wideband signal, frequency selective fading produces regions of frequencies that are severely attenuated. This process distorts the wideband signal in the time domain and causes intersymbol interference. Wideband radio systems are desirable because they avoid the problem of narrowband flat fading, however to use a wideband system the receiver must be able to deal with the distortion of frequency selective fading.

The main difference between the radio and acoustic multipath channels is the delay spread. Delay spread is the difference in time that it takes the signal to pass through the shortest path versus the time through the longest significant path. The inverse of delay spread is roughly the coherence bandwidth of the channel, which is the benchmark for defining the type of the signal. The signal is narrowband if the signal bandwidth is much less than the coherence bandwidth. It's a wideband signal if its bandwidth is much greater than the coherence bandwidth. In a small room, the acoustic delay spread may be 50ms giving a coherence bandwidth of 20Hz (the actual value varies considerably). The human voice uses a bandwidth of about 3kHz, which is much greater than the coherence bandwidth of the acoustic channel and therefore voice is wideband in its natural environment. Radio waves propagate at a much greater speed than sound so even in a large area like a cellular radio cell, the maximum delay spread may be only 50μs (again the actual value varies considerably). The coherence bandwidth of the radio channel, 20kHz in this case, is greater than the bandwidth of the voice signal, therefore voice transmitted through the radio channel behaves like a narrowband signal. This is the reason that voice carried through radio experiences flat fading, and is subject to occasional signal loss, but voice carried through the acoustic medium does not.

3. The Human Voice as a Spread Spectrum Signal

A spread spectrum signal is, by definition, a signal that occupies a much greater bandwidth than the signaling rate requires. Spread spectrum signals are most useful when the bandwidth of the signal is wide enough to avoid flat fading, while at the same time the signaling rate is low enough to avoid intersymbol interference. Modern direct sequence spread spectrum (DSSS) radios are one example of systems with this property. The human voice is another.

The wideband signal used in direct sequence spread spectrum radios is created by modulating a spreading code. The code is chosen to distribute the energy of the signal in a roughly uniform pattern across the entire frequency band so that there are no critical frequencies in the signal that could be attenuated by a frequency selective fade. As long as enough signal power falls outside the fades, the signal will get through.

The human voice is very similar to a spread spectrum radio signal. The bandwidth of the voice, about 3kHz, is much greater than the signaling rate of two to five syllables per second, and therefore the voice is a spread spectrum signal. The voice bandwidth is much greater than the coherence bandwidth of the acoustic channel to avoid flat fading, and the syllables are longer than the delay spread of a typical acoustic channel to avoid intersymbol (intersyllable?) interference. It is interesting to note that we instinctively talk slower when yelling across a room, and faster when whispering directly into someone's ear. Could this be a natural response to different delay spreads in acoustic channels?

The voice signal is, like the spreading sequences used in spread spectrum radio, inherently resistant to frequency selective fading. Three types of sounds make up a voice signal: voiced sounds, fricative sounds, and plosive sounds [2].

Voiced sounds start at the vocal cords where a quasi-periodic signal with many harmonics excites the vocal tract. The vocal tract acts as a cavity resonant filter with typically four passbands known as formants. The lips, jaw, tongue, and velum change the shape of the resonance chambers within the vocal tract, which in turn change the center frequencies of the formants to give each voiced sound it's distinct characteristic. Voiced sounds are able to pass through multipath channels because, of the many harmonics that make up each sound, none appear to be critical. In fact Miller [3], by measuring the actual vocal cord wave, demonstrated that some of the harmonics were severely attenuated or even missing at the source for certain speakers. Intonation, a variation in the fundamental frequency over time [4], may also assist the signal through the multipath channel by adding diversity to the harmonic frequencies.

Fricative sounds (such as s, f, and sh) are generated by constricting part of the vocal tract until turbulence is created. This generates a wideband noise signal similar to pseudo-noise sequences used in DSSS radios. As with voiced sounds, the signal is filtered by the vocal tract to give each fricative sound a distinct frequency response [5].

Plosive sounds (such as p, k, and t) are produced by creating a complete blockage in the airflow, and then releasing the blockage when enough pressure has built up. As wideband pulses, these sounds contain no critical frequencies that could be lost in the multipath channel.

4. The Hyperactive Chipmunk Radio

When translated to radio, the human voice does not make a good spread spectrum signal because its bandwidth is well below the coherence bandwidth of most radio channels. The hyperactive chipmunk radio compensates for this by artificially increasing the bandwidth of the voice signal through time compression. The process is shown in Figure 2a. A segment of speech is sampled and stored within the radio. When storage is complete, the voice segment is replayed at a much higher sampling rate. This compresses the signal in time and expands it in frequency. In the

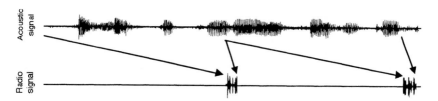

Figure 2a. The hyperactive chipmunk transmitter's segmentation and compression functions.

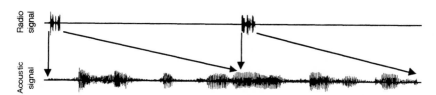

Figure 2b. The hyperactive chipmunk receiver's expansion and recombining functions.

example above, the coherence bandwidth of the radio channel is a thousand times that of the acoustic channel, so to achieve the required bandwidth expansion, the signal is played back at a sample rate a thousand times faster.

The receiver restores the signal by sampling it at the high sample rate, storing it, and playing it back at the low sample rate (Figure 2b). Not only does this restore the signal bandwidth so the listener can understand the speaker, it also expands the effective impulse response of the radio channel. If, for example, the radio channel has a delay spread of 50μs, expanding the signal by a factor of a thousand increases the delay spread to 50ms. It now appears to be a natural acoustic delay spread to the listener, which is easily handled by the aural system.

The thousandfold spreading factor used in the examples above may not be appropriate for all systems. The actual spreading factor must be chosen for the multipath environment that the radio will operate. It is desirable to spread the signal out as much as possible to avoid the possibility of flat fading. However, with too large of a spreading factor the impulse response of the radio channel may be spread out too far during expansion. This makes the multipath echoes noticeable to the listener, as if the conversation were held in a large empty room or cave.

Single sideband (SSB) modulation is the preferred method for modulating hyperactive chipmunk signals onto the radio carrier. It is a linear modulation technique, so the multipath characteristics of the radio channel are preserved when translated to baseband, and is bandwidth efficient. Double sideband (DSB) modulation is another linear modulation scheme that can be used, but with only half the bandwidth efficiency. DSB modulation also results in a poorer signal to noise ratio than SSB for a given transmit power. This is different from the common narrowband case where SSB and DSB modulation produce the same SNR [6]. The difference comes in the way that the upper and lower sidebands combine within the DSB receiver. With narrowband systems, the upper and lower sidebands fall within the coherence bandwidth of the radio channel and combine coherently within the receiver. In this wideband system, the sidebands are separated by more than the coherence bandwidth of the channel and must be treated as independent stochastic signals and combined non-coherently. Non-coherent combining results in only about half the signal energy of coherent combining therefore wideband DSB receivers suffer a 3dB SNR penalty. Wideband SSB receivers do not combine sidebands and are not subject to this penalty. Vestigial sideband modulation is another effective modulation technique, however it suffers the same bandwidth and SNR penalties as DSB, to a lesser degree. Amplitude modulation is not suitable because it contains one critical frequency, the carrier tone, that is susceptible to frequency selective fades. Nonlinear modulation techniques such as frequency and phase modulation do not preserve the multipath distortion characteristics of the channel and sound unnatural.

One further consideration is the edge effects that are generated when a segmented voice signal is sent through a multipath channel. Figure 3a demonstrates the process by using a rectangle to represent a voice segment of length T. The delay spread of the channel causes the signal to spread out by T_d seconds so there is a part of the signal missing near the beginning of the received segment and an extra signal tail added to the end. The edge effects reduce the usable part of the voice segment to $T-T_d$ seconds. To overcome this problem, a precursor is added to the beginning of each voice segment as shown in Figure 3b. The precursor consists of the last T_d seconds of the previous voice segment. Although the precursor is discarded at the receiver, it does move the edge effect away from the beginning of the true voice segment. The useful part of

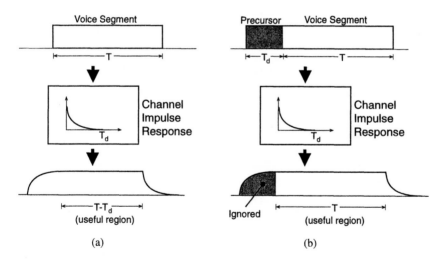

Figure 3. The effect of delay spread on the compressed voice segments (a) without a precursor, and (b) with a precursor.

the voice segment is now T seconds long, which is enough to be reassembled back into a continuous voice signal.

5. Advantages and Disadvantages of the System

Complexity

The major advantage of this system compared to other spread spectrum systems is its simplicity. No correlators, equalizers, or RAKE receivers are required to handle the multipath channel distortion. These functions are performed within the listener's ear and brain. The time compression and expansion circuitry can be implemented with sample/hold devices and charge coupled device (CCD) arrays, which eliminates the need to digitize the signal.

Compared to narrowband analog systems, this system is more complex, mainly due to the synchronization required to capture the time compressed segments. However, if there is a digital control channel associated with the system, then the analog synchronization can be tied to the digital synchronization with no additional cost.

Bandwidth Efficiency

Despite being a spread spectrum system, the bandwidth efficiency of the hyperactive chipmunk radio is very high. For example with SSB modulation and a thousandfold compression factor, the signal will occupy a 4MHz radio bandwidth (including guardbands). Allowing for a pessimistic one third of the system time for the overhead associated with precursors, synchronization, and control signals, it is possible to time multiplex 666 voice channels within that bandwidth. This results in an efficiency of 166 one way voice channels per megahertz of bandwidth, five times the AMPS cellular phone standard.

Power

When compared to a narrowband system, the average power requirement for a hyperactive chipmunk radio is substantially lower. Extra power is normally added to narrowband signals to allow them to pass through all but the deepest flat fades. This fading margin may add 20dB or more to the output power at the transmitter. The wideband signals of the hyperactive chipmunk radio, like other spread spectrum radios, are not as susceptible to flat fading and the fading margin can be virtually eliminated.

Although the average output power is lower than the power from a narrowband radio, the power amplifier in a hyperactive chipmunk radio must actually be larger. This is a result of the small duty cycle of the transmitted signal. For example, to transmit an average power of 10mW with a compression ratio of 1000, the radio actually transmits 10W with a 0.1% duty cycle. A linear RF power amplifier is required that can transmit a relatively large RF power and then quickly shut down between voice segments.

Narrowband Interference

Like other spread spectrum radios, the hyperactive chipmunk radio naturally attenuates narrowband interference. The mechanism that achieves this signal rejection is time windowing. The receiver samples the incoming radio signal only in the short periods of time when the desired voice segment arrives. Between the voice segments, the signal is ignored. For a compression factor of one thousand, only 0.1% of the interfering signal's energy arrives during the time that the receiver is sampling, therefore its effect on the restored signal is reduced by a factor of 30dB.

Delay

The process of compressing the voice signal in the transmitter introduces a significant delay in the system. As shown in Figure 2a, the delay is equal to the length of the uncompressed voice segments. To keep the two-way delay reasonable for a normal conversation, the voice segment length should be less than 100ms. However, making the segment length too short causes bandwidth efficiency to suffer as the precursor consumes a larger percentage of transmission time.

Doppler

The hyperactive chipmunk radio works best when the radio channel is relatively stable. High Doppler situations caused by high speeds and/or high carrier frequencies cause the radio channel to change considerably between voice segments and create discontinuities at the boundaries when the segments are reassembled. The discontinuities caused by minor Doppler can be reduced by overlapping voice segments slightly and providing a gentle transition from one segment to the other.

Carrier Frequency Recovery

Single sideband modulation requires very accurate carrier frequency at the receiver, within a few hertz, to clearly demodulate a narrowband voice signal. This usually necessitates some kind of a carrier recovery circuit. The wideband signals used in the hyperactive chipmunk radio and the expansion operation in the receiver desensitize the system so offsets of several kilohertz can be tolerated. Practical free running oscillators can generate radio carriers within this tolerance, so no carrier recovery circuit is needed.

6. Conclusions

The human voice makes an effective spread spectrum signal that can be applied to the radio domain as well as its natural acoustic domain. The result is a very simple implementation of a spread spectrum radio system. The system is able to operate at a lower power level than narrowband analog systems, with little chance of signal loss from flat fading. Its bandwidth efficiency is very high, and it naturally attenuates narrowband interference. The major technological challenge is the requirement for a relatively high power, linear amplifier that is able to handle the short bursty signal.

Acknowledgements

This project is funded by TRLabs (Telecommunications Research Laboratories), a western Canada based consortium of government, university, and industrial sponsors. The authors would like to thank internship students Alex Kavanagh and Ryan Schneider for their work in verifying many the concepts presented in this paper by computer simulation and laboratory experiment. Appreciation is also extended to the many "guinea pigs", who patiently listened to voice clip tests and offered their opinions.

References

[1] J.G. Proakis, "Digital Communications", second edition, McGraw-Hill, 1989.

[2] J.L. Flanagan, "Voices of Men and Machines", J. Acoustic Soc. Am., vol. 51, pp. 1375-1387, March 1972.

[3] R.L. Miller, "Nature of the Vocal Cord Wave", J. Acoustic Soc. Am., vol. 31, pp. 667-677, June 1959.

[4] G. Fant, "The Acoustics of Speech", Proc. Third International Congress on Acoustics, pp. 188-201, 1959.

[5] J.M. Heinz and K.N. Stevens, "On the Properties of Voiceless Fricative Consonants", J. Acoustic Soc. Am., vol. 33, pp. 589-596, Man 1961.

[6] A.B. Carlson, "Communication Systems: An Introduction to Signals and Noise in Electrical Communication", third edition, McGraw-Hill, 1986.

Turbo Code Implementations on Fixed Point DSP's

Eric Cress, Student Member IEEE
eic@cypress.com
and
William J. Ebel, Member IEEE
ebel@ece.msstate.edu
Mississippi State University
Electrical and Computer Engineering
Box 9571 Mississippi State, MS, 39762 USA

Abstract

This paper presents the results of research in the implementation of turbo decoding algorithms on the TI TMS320 digital signal processor. Empirical results are presented for algorithms that were published in [2]. This study focuses on implementing turbo decoding algorithms on the TMS3206201 fixed point DSP architecture, but the issues that are discussed are applicable to other processing architectures such as FPGA, CPLD, and ASIC. Benchmarks for evaluating the performance of hardware implementations are featured, as well as performance results for efficient implementations on the Texas Instruments TMS320C6201 fixed point DSP.

1 Introduction

There is much excitement in the communication industry over the relatively new theoretical class of forward error correcting codes known as Turbo Codes [1]. It has been shown that these codes approach the Shannon limit for reliability improvement on an AWGN channel. However, the question remains, is this class of codes practical from a hardware standpoint? What are the hardware complexity issues, and what kind of bit-rate can one expect from a Turbo Decoder? In this paper we address this question by presenting implementation options and performance results for a Turbo Decoder implemented on the TMS320C6201 fixed-point digital signal processor.

This Turbo Decoder implementation is parallel concatenated code. The encoder is shown in Figure 1. Figure 2 illustrates the rate $\frac{1}{2}$ convolutional encoder used for the constituent encoders in the Turbo Encoder, which is rate $\frac{1}{3}$. Precursor bits were used to ensure trellis termination, as discussed in [2].

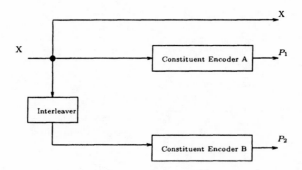

Figure 1: Turbo Encoder

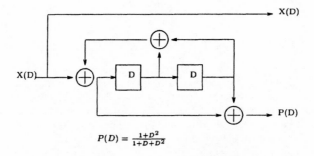

$$P(D) = \frac{1+D^2}{1+D+D^2}$$

Figure 2: Convolutional Encoder

The encoder can be viewed in the form of a state diagram. Figure 3 shows the state diagram for the encoder shown in figure 2, where the transition between four distinct states is represented by the branches between the states. The branch taken in the transition between states is dependent on the information input, x. The output of the decoder is the information input,x, and a parity bit, p. At the receiver end, the decoder determines the probability that a certain series of information bits was sent by finding the probability that specific states and transition branches were traversed, given the received information and parity signals. As a conceptual aid, it helps to look at the state diagram as a trellis. The trellis simply *stretches* the state diagram over time, so that the transition probabilities and state probabilities may be recorded for each stage of the information sequence. The trellis for the encoder is shown in figure 4.

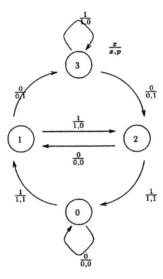

Figure 3: State Diagram Representation of Convolutional Encoder

The probability that a given state transition branch was taken given the received sequence at a given stage is defined as the γ metric. The probability that a given state was traversed given all previously received bits is defined as the α metric. The probability that a given state was traversed given all subsequent received bits (until the end of the trellis) is defined as the β metric. Both the α and β metrics are recursively computed using only the known initial starting state (or known initial ending state for the β metrics), and the branch metrics.

Once the metrics are computed for each stage, the probability that a given branch was taken given the entire received sequence may be computed. This calculation is the product of the γ metric for the branch, the α metric for the

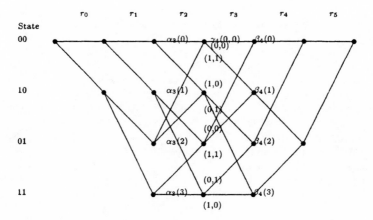

Figure 4: Trellis representation of convolutional encoder/decoder

state that the branch is leaving, and the β metric for the state to which the branch is directed. The *likelihood ratio* for the probability that a binary 1 was sent versus a binary 0 is computed by summing all of the branch metrics at a given stage that represent a binary 1 and dividing that sum by the sum of all the branch metrics at the stage that represent a binary 0. These relationships are stated mathematically by the following four relations.

$$\alpha_{i-1} \cdot \gamma_i \cdot \beta_i = P\{r_1^{i-1}|s_{i-1}\} \cdot P\{r_i, s_{i-1}^i\} \cdot P\{r_{i+1}^N|s_i\} \tag{1}$$

$$\alpha_i = \frac{\sum \alpha_{i-1} \cdot \gamma_i}{P\{s_i\}} \forall s_{i-1} \qquad \text{where } \gamma_i(s_{i-1}^i) \text{ exists.} \tag{2}$$

$$\beta_i = \frac{\sum \beta_{i+1} \cdot \gamma_{i+1}}{P\{s_i\}} \forall s_{i+1} \qquad \text{where } \gamma_i s_i^{i+1} \text{ exists.} \tag{3}$$

$$\gamma_i = P\{r_i, s_{i-1}^i\} \tag{4}$$

$$= P\{r_i|s_{i-1}^i\} \cdot P\{s_{i-1}^i\} \tag{5}$$

$$\tag{6}$$

The above relations are greatly simplified if we use *log-likelihood ratios* as opposed to likelihood ratios. By taking the natural log of the metrics, we can transform the multiplication and division in the computations to addition and subtraction. Using log-likelihood ratios, and simplifying the calculation algebraicly, the alpha's, beta's, and gamma's become the following.

$$A_i = ln(\sum(A_{i-1} + \Gamma_i)) \qquad \text{where } \gamma_i(s_{i-1}^i) \text{ exists.} \tag{7}$$

$$B_i = ln(\sum(B_i + \Gamma_i)) \qquad \text{where } \gamma_i(s_{i-1}^i) \text{ exists.} \tag{8}$$

$$\Gamma_i = \sum_{m=0}^{M-1} \frac{2r_{i,m}y_{i,m}}{\sigma^2} \cdot b_{i,m} \tag{9}$$

where the subscript m denotes the component of the received symbol at a given stage [1], and the subscript i represents the stage of the trellis. x_i and y_i represent the binary value and modulated value that correspond to the given branch, respectively [2].

The log-likelihood that a binary 1 information symbol was sent at a given stage versus a binary 0 is given as $\Lambda(X)$.

$$\Lambda(X) = ln[\frac{\sum(A_{i-1} + \Gamma_i + B_i) \quad \forall(y_{i,m} = b)\exists s_{i-1}^i}{\sum(A_{i-1} + \Gamma_i + B_i) \quad \forall(y_{i,m} = -b)\exists s_{i-1}^i}] \tag{10}$$

The communications model was taken to be of the form shown in figure 5. The modulator is binary, and the channel noise is taken to be AWGN, and is implemented by adding a Gaussian random variable to each transmitted symbol. All operations inside the dotted box were modeled in software using a custom C++ simulation environment.

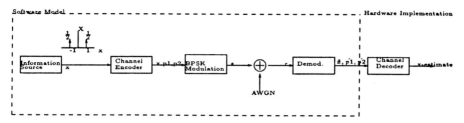

Figure 5: Communication Model

The general Turbo Decoder architecture that was implemented is illustrated in figure 6.

2 Implementation

2.1 Development and Test Environment

The hardware and software environment used to implement the Turbo Decoder consisted of an evaluation module provided by Texas Instruments and a software development environment provided by GO DSP. All hardware used in the development was on the TMS320C6201 EVM (evaluation module). The primary

---

[1] for the rate $\frac{1}{2}$ encoder $r_{i,0}$ represents the received signal for x_i and $r_{i,1}$ represents the received signal for p_i

[2] For a branch that represents $x_i = 1$ and $p_i = 0$, in the equation $b_{i,0} = 1$, $y_{i,0} = 1$, $b_{i,1} = 0$, and $y_{i,1} = -1$

188

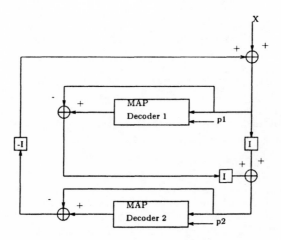

Figure 6: General Turbo Decoder Architecture

hardware resources that were used in the decoder implementation are listed as follows.

- TMS32062001 fixed point digital signal processor with a maximum clock frequency of 200MHz.

- 64k bytes of internal program memory running at system clock speed.

- 64k bytes of internal data memory running at system clock speed.

- 8M bytes of SDRAM located on the EVM running at 100MHz maximum clock frequency ($\frac{systemclockspeed}{2}$). The SDRAM was used to store the sample data and post processing of the decoded data to gather error statistics.

The limited memory resources of the hardware and the desire for high throughput demand the judicious use of memory and computational resources. Fortunately, careful storage and memory re-use allowed for maximum throughput for block lengths up to 2000 information bits using only the DSP's internal data memory for storage of all computation. For applications requiring more memory, the external SDRAM could be used to allow block sizes as long as 512,000 information bits.

2.2 Memory Organization

All memory resources are accessed via the TMS320's on-board DMA controller. The TMS320C6x compiler allows flexible mapping of the memory resources. A memory model is first selected in order to divide the memory up into regions that characterize the size and speed of the memory. The memory model used for the Turbo Decoder implementation is shown in table 1. The fastest memory regions are the internal program memory (IPM), and internal data memory (IDM). The IPM stores the actual DSP executable code. The IDM stores the stack, local variables, and any variables that require high-performance memory. The remaining memory regions are the slower external memory. Table 2 shows

Memory Type	Origin (Hex)	Length (Hex Bytes)	Length (Hex Bytes)	Type
INTPROG	0x00000000	0x010000	64k	IPM
INTDATA	0x80000000	0x010000	64k	IDM
EXTMEM0	0x00400000	0x040000	256k	SBSRAM
EXTMEM1	0x02000000	0x400000	4M	SDRAM
EXTMEM2	0x03000000	0x400000	4M	SDRAM

Table 1: Memory Model Used for the Turbo Decoder Implementation

how the various program variables are assigned to memory regions. Most of the region assignments are fairly general. However, the decoder working memory, sample data, and error statistics section assignments were made due to the program's requirement for performance during certain variable accesses. The Turbo Decoder's working memory is assigned to the IDM for high performance, while all post-processing memory is assigned to the slower memory regions.

Region	Variable	Description	Type	Size
INTDATA	IMAP[BS]	Interleaver map	ushort	$2 \cdot BS$
INTDATA	IMAPU[BS]	Deinterleave map	ushort	$2 \cdot BS$
INTDATA	LxAD[BS]	received x sample	short	$2 \cdot BS$
INTDATA	Lp1AD[BS]	received parity 1 sample	short	$2 \cdot BS$
INTDATA	Lp2AD[BS]	received parity 2 sample	short	$2 \cdot BS$
INTDATA	Lext1[BS]	MAP dec. 1 extrinsic data	short	$2 \cdot BS$
INTDATA	Lext2[BS]	MAP dec. 2 extrinsic data	short	$2 \cdot BS$
INTDATA	A[BS][NS]	Alpha calculations	short	$8 \cdot BS$
INTDATA	B[BS][NS]	Beta calculations	short	$8 \cdot BS$
INTDATA	numErrors	Post-processing data	unsigned	4
			Total:	$30 \cdot BS$
EXTMEM1	xSamples[BS · NBL]	all received x samples	short	$2 \cdot BS \cdot NBL$
EXTMEM1	p1Samples[BS · NBL]	all received parity 1 samples	short	$2 \cdot BS \cdot NBL$
EXTMEM2	p2Samples[BS · NBL]	all received parity 2 samples	short	$2 \cdot BS \cdot NBL$
EXTMEM2	xOut[$\frac{BS \cdot NBL}{16}$]	binary x estimate from decoder	short	$\frac{BS \cdot NBL}{8}$
EXTMEM2	goldData[$\frac{BS \cdot NBL}{16}$]	source x data from encoder	short	$\frac{BS \cdot NBL}{8}$
			Total:	$6\frac{1}{4} \cdot NBL \cdot BS$

BS = interleaver size
NS = Number of encoder states
NB = Number of branches from a state
NBL = Number of blocks

Table 2: Decoder Memory Section Assignments

Sequence		Likelihood Ratio	Log-Likelihood Ratio
X	P		
0	0	$e^0 \cdot e^0 = 1$	$ln(1) + ln(1) = 0$
0	1	$e^0 \cdot e^{\frac{2\cdot p}{\sigma^2}} = e^{\frac{2\cdot p}{\sigma^2}}$	$ln(1) + ln(e^{\frac{2\cdot p}{\sigma^2}}) = \frac{2\cdot p}{\sigma^2}$
1	0	$e^{\frac{2\cdot x}{\sigma^2}} \cdot e^0 = e^{\frac{2\cdot x}{\sigma^2}}$	$ln(e^{\frac{2\cdot x}{\sigma^2}}) + ln(1) = \frac{2\cdot x}{\sigma^2}$
1	1	$e^{\frac{2\cdot x}{\sigma^2}} \cdot e^{\frac{2\cdot p}{\sigma^2}}$	$ln(e^{\frac{2\cdot x}{\sigma^2}}) + ln(e^{\frac{2\cdot p}{\sigma^2}}) = \frac{2\cdot x}{\sigma^2} + \frac{2\cdot p}{\sigma^2}$

Table 3: γ calculations for received symbols

2.3 Computation

The computational complexity of the Turbo Decoder is dominated by the MAP decoder implementation; namely, the alpha, beta, and gamma metrics must be computed for every stage in the block. Therefore, the performance of these three computations limit the speed of the decoder.

2.3.1 Gamma Computation

the gamma metric stores the result of the computation given in equation 9. For the $\frac{1+D^2}{1+D+D^2}$ encoder that was chosen for this implementation, there are 4 possible received symbol sequences. These sequences are shown in table 3. Because the TMS320C6201 has a 4 cycle memory fetch, and to save valuable internal data memory resources, it was decided that incorporating the gamma calculations into the alpha and beta computation routines was the most efficient implementation. Table 3 shows that the gamma calculation actually only requires one computation. Redundantly calculating the one computation turns out to be more efficient than storing the data to memory. The results in section 3.1 give empirical data that back this decision.

2.3.2 Alpha/Beta Computation

The alpha and beta computation routines implement the computation given in equations 7 and 8, respectively. The summation that is given in equation 7 represents the addition of the branch metrics for multiple branches entering a given state. Taking the natural log of this sum is computationally expensive. The simple approximation is made to circumvent the natural log computation. If we assume that in most cases $A \gg B$, then we can approximate the exponential adder as just a magnitude comparison:

$$ln(e^A + e^B) \approx \text{MAX}(A, B) \quad \forall A \gg B \tag{11}$$

This approximation has been shown in [2] to yield good performance, with a slight coding loss. Implementing the LOG MAP decoder, which does not use the approximation shown in equation 11, would require a look-up table which would be $16 \cdot 2^{16}$ in size to match the decoder's precision, or suffer some loss due to approximation for a smaller size table. Information on the performance

Method	clocks/MAP decode	bps/iteration
Dedicated gamma calculation	136,700	88,000
Integrated gamma calculation	61,268	196,000
Integrated gamma/flat arrays	42,000	286,000

Table 4: Bit-rate performance results for different implementations.

degradation due to the LOG MAP MAX decoder which uses the approximation in equation 11 can be found in [2] and [4]

The following shows the forward recursive calculation of one alpha metric at a given stage, taking advantage of the TMS3206X's _sadd intrinsic which computes a saturated add. The beta calculations are similar, except the recursion is backwards, which changes the gamma metrics' assignment to the state metrics.

```
/** this is our one gamma calculation */
g = _sadd(Lest[iStage-1]<<16, Lp[iStage-1]<<16);

/** i1 is one branch, i2 is the second branch entering a state */
i1 = *(Aptr+jStageAlpha-4+0)<<16;
i2 = _sadd(*(Aptr+jStageAlpha-4+2)<<16, g); /* g21 */
/** our approximated ln[exp(i1) + exp(i)] */
*(Aptr+j+0) = (short)(_VMAX(i1, i2)>>16);  /*~ ln[exp(i1) + exp(i)] */
```

2.3.3 Information LLR Calculation

Calculation of the log-likelihood ratios involves the implementation of equation 10. The approximation for the log-add given in equation 11 is again utilized here. The following code snippet illustrates the calculation of the $x = 1$ log-likelihood ratio from the alpha's and beta's (Aptr and Bptr, respectively).

```
/** Our one gamma calculation */
g = Lp[i]<<16; /*g21, g01*/
/* calc 1's LL */
LL1 = _VMAX(_sadd(_sadd(*(Aptr+j+0)<<16,g),*(Bptr+j+1)<<16),
            _sadd(*(Aptr+j+1)<<16,*(Bptr+j+2)<<16));
LL1 = _VMAX(LL1, _sadd(_sadd(*(Aptr+j+2)<<16,g),*(Bptr+j+0)<<16));
LL1 = _VMAX(LL1, _sadd(*(Aptr+j+3)<<16,*(Bptr+j+3)<<16));
```

3 Performance Results

The performance of the decoder was measured in bit-rate and coding gain. The coding gain of the decoder matches the statistics discussed in [2] and are not discussed here. In this section we give the test results of the bit-rate for the implementation.

3.1 Bit-Rate

Several strategies were employed to increase the overall bit-rate of the decoder. A few of these strategies were particularly successful in increasing the overall throughput. Table 4 gives bit-rate performance for several strategies.

3.1.1 Integrated gamma calculation

A method was presented that involved the calculation of the branch probabilities, also referred to as the gamma calculations. In the dedicated gamma calculation method, functions independently calculate all of the gamma calculations for the entire trellis, and store them to memory. As shown in table 3, the gamma calculation actually only requires one computation, the rest of the operation is a simple reassignment of variables. Due to the high cost of memory accesses a separate gamma calculation is very inefficient. Alternately, the gamma calculation can be pulled into each of the alpha and beta calculation functions. This saves the processor numerous memory accesses per iteration, at the cost of only two redundant computations (the branch computation in which the information and parity bit are both one must be done in the alpha calculation, the beta calculation, and information LLR calculation independently). As shown in table 4, the integration of these calculations leads to a considerable performance improvement of over 100% increase in bit-rate.

3.1.2 Integrated gamma/flat arrays

The variables that store the alpha, beta, and final information estimate are each multi-dimensional arrays. It was discovered that the compiler is not very efficient in computing the pointer addresses for the array accesses. Since the array accesses are sequential in nature, the compiler can be helped along if the programmer *flattens* the arrays into one-dimension and manually increments the indexes rather than relying on the compiler to make efficient computations. This enhancement further improves the bit-rate by about 33%.

4 Conclusion

The purpose of this paper is to present the results of investigations in the implementation of Turbo-Decoding algorithms on hardware architectures. We show that on the TMS320C6201 fixed-point DSP, a common DSP architecture, that decoders for codes of block sizes on the order of 2000 information bits can be implemented just using the DSP alone. Much larger block sizes can be implemented using peripheral memory. In fact, decoders with block sizes on the order of 512,000 information bits can be implemented on the EVM architecture. For decoders implemented on the DSP's internal memory space, bit-rate performance as high as 286,000 bits/second/iteration can be achieved. Though the C-code for this specific implementation takes advantage of the TMS320C6x's specific architecture, implementations that are written in processor-specific assembly code should be able to achieve bit-rates that are significantly higher. The authors feel that as DSP technology progresses, and particularly for ASIC designs, bit-rates on the order of Mega-bits per second are achievable, and should be expected.

References

[1] Berrou, C., Glavieux, A., and Thitimajshima, P., "Near Shannon Limit Error-Correcting Coding and Decoding: Turbo-Codes (1)", International Communications Conference, Geneva, Switzerland, 1993, pp. 1064-1070.

[2] D. E. Cress, W. J. Ebel, "Turbo Code Implementation Issues for Low Latency, Low Power Applications". *1998 Symposium on Wireless Personal Communications*, MPRG, Virginia Tech, June 10-12, 1998.

[3] W. J. Ebel, "Turbo Codes: Algorithms and Performance". Final Report, Contract 305425-060700-021000, Texas Instruments, December 1996.

[4] Viterbi, Andrew J., "An Intuitive Justification and a Simplified Implementation of the MAP Decoder for Convolutional Codes", *IEEE Journal on Selected Areas In Communications*, February 1998, pp. 260-264.

TCP with Adaptive Radio Link

Dongjie Huang and James J. Shi

Ericsson Inc.

740 East Campbell Road MP-9, Richardson, TX 75081, USA

E-mail: dongjie.huang{james.shi}@ericsson.com

Abstract

The performance of TCP over wireless links could degrade due to handoff, high bit error rate and long roundtrip delay on the air interface. The conventional Radio Link Protocol uses fixed channel coding and ARQ to mitigate impairment over wireless channels. Adaptive channel coding with the use of punctured convolutional codes has been proposed for its ability to adapt to channel quality and maintain a high level of system throughput. In this paper, we study TCP over adaptive RLP based on estimated channel condition. We consider a Markov channel model and evaluate the performance of TCP with adaptive RLP in such environment. We also evaluate the existing channel quality measurement in wireless systems and propose a methodology to update RLP channel coding schemes based on filtered channel measurement. We use simulation to calculate the throughput of TCP system and determine the effectiveness of adaptive channel coding with respect to different fading rates.

1 Introduction

Transmission Control Protocol (TCP) has been successfully used in wireline data networks to ensure end-to-end performance. TCP was designed for traditional wireline networks where congestion contributes to most of packet loss and unusual delay [1]. The performance of TCP over wireless links could degrade due to handoff, high bit error rate and long roundtrip delay on the air interface. The congestion control measures developed for wireline networks would cause unnecessary reduction of throughput in wireless networks. Radio Link Protocols (RLPs) have been proposed to alleviate the effect of non-congestion-related loss over wireless links [2] [3].

The conventional RLP uses a fixed channel coding scheme and automatic retransmission request (ARQ) to mitigate impairment over wireless channels. The throughput of the RLP depends on the rate of channel coding and rate of retransmission due to error [4]. Adaptive channel coding [5] [6] with the use of punctured convolutional codes has been proposed for its ability to adapt to channel quality and maintain a high level of system throughput.

It has been shown in [7] that RLP with adaptive channel coding and ARQ may significantly improve TCP performance over fixed channel coding. However, it was assumed that the system has perfect knowledge of channel condition and updates its coding scheme instantaneously according to channel condition. In this paper, we study TCP over adaptive RLP based on estimated channel condition. First, we consider a 2-state Markov channel model for the evaluation of TCP performance with adaptive RLP. Second, we discuss the existing channel quality feedback mechanism in wireless systems and propose a methodology to update RLP coding schemes based on channel quality feedback. Third, we combine the channel model and channel quality feedback and use simulation to calculate the throughput of TCP system. We compare the performance of such system to the ideal performance obtained in [7] for different fading rates.

The rest of this paper is organized as follows. Section 2 reviews basic concepts of adaptive channel coding and ARQ, and describes a data communications model and a Markov channel model. Section 3 proposes a methodology to update channel coding based on estimated channel condition. Section 4 presents simulation results of TCP over adaptive RLP. Finally, Section 5 concludes the paper.

2 Basic Concepts

In this section, we first discuss application of punctured convolutional codes and ARQ for adaptive RLP and their performance criteria. We then describe a wireless data communications model for TCP over RLP. Finally, we discuss a 2-state Markov channel model.

2.1 Adaptive Channel Coding and ARQ

A regular (n, k, m) convolutional code can be implemented with k-input, n-output linear sequential circuit with input memory m. Typically, n and k are small integers with $k < n$, but the memory order m must be made large enough to achieve low error probability. The rate of the convolutional code is $R = k / n$. The performance of a convolutional code depends on the minimum free distance d_{free}, which is defined as the minimum distance between any two code words. A performance bound of convolutional codes can be derived from the minimum free distance. In [4], the probability of bit error after decoding is bounded by

$$P_b(E) < \frac{1}{k} \sum_{d=d_{free}}^{\infty} B_d P_d \tag{1}$$

where B_d is the total number of nonzero information bits on all weight d paths and P_d is the probability of first event error of an incorrect path with weight d and is approximated by

$$P_d \approx \begin{cases} \binom{d}{(d+1)/2} p^{(d+1)/2} (1-p)^{(d-1)/2}, & d \text{ odd} \\ \frac{1}{2} \binom{d}{d/2} p^{d/2} (1-p)^{d/2}, & d \text{ even}. \end{cases} \tag{2}$$

where $p \ll 1$ is the raw bit error rate (BER) of the input sequence, referred to as BER in this paper. Note that both d_{free} and B_d can be obtained by searching all code words.

The minimum free distance represents the error correction capability of the convolutional code, and the code rate represents the code efficiency. One way to increase the code efficiency is to puncture a low rate code. Punctured convolutional [6] have the advantage of changing the code rate without changing the basic structure. In other words, the same encoder and decoder can be used for a set of punctured codes derived from the same basic code. As an example, a rate 5/6 punctured code can be obtained from the basic rate 1/2 code by choosing only every fifth bit of the first output. Similarly, a rate 4/5 punctured code can be obtained by choosing every fourth bit of the first output, and so on. It has been shown that punctured codes are as good as the best known codes and some of the good punctured codes use the same basic rate 1/2 generators. These codes have been used in many RLP standards [2].

ARQ is frequently used in data communications protocols. When a frame is detected to contain errors after decoding, it is discarded and an acknowledgment is sent back to the sender requesting a retransmission of the frame. This is called selective retransmission and is an efficient way of retransmission. We only consider ARQ with selective retransmission.

Adaptive channel coding with ARQ has received much attention. There are two methods to adapt channel coding to channel condition. The first method starts with the channel code that has the least redundancy and transmits more redundant bits and/or retransmits as needed until the information can be decoded error free at the other end [5]. The second method changes the channel coding used in both transmission and retransmission based on the current channel condition [7]. The punctured convolutional codes fit well with the first method and can be used in the second method too. We consider the second method in our simulation.

Probability of retransmission is the probability that the frame contains at least one error. In practice, the latter is often called frame error rate (FER). FER can be calculated from $P_b(E)$ of the convolutionally decoded frame. Assuming a frame length of l, we have

$$FER = 1 - (1 - P_b(E))^l . \tag{3}$$

It follows that the relative throughput (T_r) of a data communications system using a rate-R convolutional code and selective ARQ can be written as

$$T_r = R(1 - FER) = R(1 - P_b(E))^l . \tag{4}$$

2.2 TCP Over RLP

A simplified model of protocol stack for circuit-switch data communication has been described in [7] and is shown in Fig.1. Here TCP packets from an end user are first added with IP and PPP related overhead and then forwarded to RLP layer. RLP layer adds its own overhead and then forwards it to the physical layer supported by the IS-136. For this model, we assume a high-speed link from MSC to TCP server and the Round Trip Time (RTT) between MSC and TCP server is negligible when compared to the RTT spent on the wireless link. The TCP, IP and PPP protocols in the stack are essentially the same as those in the regular wireline network. The PPP protocol generates relatively fixed overhead to the system performance and is not considered in the study. The RLP layer below PPP layer uses selective repeat, error correction and error detection to provide reliable data transmission. Additionally, the proposed adaptive RLP is enhanced with adaptive channel coding based on wireless channel condition. The physical layer below RLP uses digital traffic channel supported by IS-136 air interface.

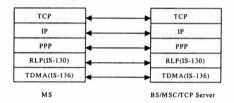

Fig. 1. Model for TCP/IP over Wireless Link.

2.3 Markov Channel Model

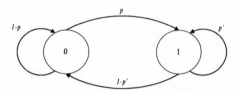

Fig. 2. Two state Markov channel model.

We consider a two state Markov model [8] for representing a wireless channel. In Fig. 2, state 0 represents the quiet state in which the BER (ε_0) is low and single bit errors are produced independent of other errors. State 1 represents the noisy or burst state, where the BER (ε_1) is high. The following parameters are derived from the model, the average burst length (b), the average BER ($\bar{\varepsilon}$) and the duty cycle (d).

$$b = \frac{1}{1 - p'}$$

$$\bar{\varepsilon} = \frac{(1 - p')\varepsilon_0 + p\varepsilon_1}{(1 - p') + p}$$

$$d = \frac{p}{(1 - p') + p}$$

As shown in [8], we take $\varepsilon_0 = d\bar{\varepsilon}$. If b, $\bar{\varepsilon}$ and d are defined, we can calculate all the other parameters of the model using the four equations. In this work, we discuss the case of burst channel which has an average BER ($\bar{\varepsilon}$) of 1.8% and a duty cycle (d) of 0.28. This translates to $\varepsilon_0 = 0.5\%$ and $\varepsilon_1 = 5\%$. The same analysis can be done for other kinds of channels. We will vary the average burst length to characterize different fading rates. The minimum time for the channel to stay in the same state is determined by the data frame duration (for example, 20 ms for the IS-136 standard).

3 Adaptive RLP

For analysis, we consider a set of commonly used punctured convolutional codes derived from the same rate 1/2 convolutional code. They are rate 5/6, 4/5, 3/4, 2/3, and 1/2, respectively. For comparison, we also include the uncoded case. The length of the frame is 260 bits, taken from the IS-130 standard. Fig. 3 shows the throughput of these codes with respect to the BER. It is seen that each code, except for the rate 3/4 code, offers optimum performance for a certain range of BER. Considering the complexity, the combination of rate 5/6 and 1/2 codes provides near optimum performance for the entire range of BER. The corresponding adaptive algorithm is to use rate 5/6 code for BER below 0.028 and rate 1/2 code for BER above 0.028.

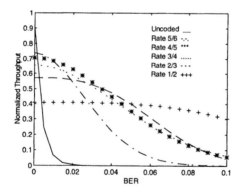

Fig. 3. Throughput with respect to the BER.

Current generation wireless systems monitors channel condition through channel quality feedback mechanism. In general, Channel feedback includes received signal strength, BER and FER. As an example, the IS-136 standard defines a channel quality measurement scheme in which the mobile station calculates bit errors in every frame (40 ms, equivalent to two time slots or two data frames) and reports back to the base station the moving average of BER every second (25 frames).

200

By slightly modifying the existing scheme, it is possible to send back the bit error information to the base station every 20 ms. This allows the base station to estimate channel condition and change its channel coding scheme every 20 ms if necessary. To reduce the overhead introduced by changing the coding scheme, similar to what is performed by the mobile station in the IS-136 standard an averaging operation can be performed in the base station on the reported bit error information. This will eliminate those unnecessary changes of coding scheme due to fading that lasts only a small duration such as a frame length. The base station uses the estimated channel condition to determine the appropriate coding scheme. We will use the above scheme in the simulation.

4 Simulation Results

The simulation is carried out based on the model shown in Fig. 1. The TCP layer algorithms by Jacobson [9] and Karn [10] for Internet congestion control that have major impact on the system throughput are implemented in the simulation. The simulation focuses on the TCP behavior under dynamic radio environment and the throughput for the TCP link in an established state. We assume that TCP maximum segment size is 536 bytes and sender's TCP window is limited by congestion window size and not by receiver advertised window size. At the RLP layer, the transmission mechanism of RLP1 of IS-130 is adopted. The forward error correction is modified with adaptive coding schemes such that the rate 5/6 code is used in state 0 of the two state Markov channel model in Fig. 2 and the rate 1/2 code is used in state 1. Undetectable errors are not considered in this study since it is very unlikely with the specified CRC in IS-130. The physical layer uses a full rate digital traffic channel, which could carry 260 bits of data for every 20 ms. The channel fading rate of radio environment in our study is represented by the average burst length in the two state Markov channel model. The average burst lengths used in simulation are 20 ms, 100 ms and 500 ms, which are equivalent to 1, 5 and 25 data frames of IS-130 respectively. Other parameters for the channel model are specified in Section 2.3.

TCP throughput is given for the above three types of fading rates in this section. They provide the TCP performance response to the continuous traffic on the forward channel from the base station to the mobile station. Under each fading rate, TCP throughputs are obtained for ideal adaptive channel coding, adaptive channel coding based on channel estimation, fixed rate 5/6 and fixed rate 1/2 coding. In an ideal adaptive channel coding approach, a transmitter knows channel condition beforehand and could therefore choose the best coding scheme accordingly. The adaptive channel coding based on channel estimation would choose a coding scheme based on the history of smooth averaging BER. The two fixed rate schemes use the same channel coding throughout the data transmission. They apply to the case that channel coding is selected at the time of connection but maintained during the transmission. Figs. 4-6 compare the TCP performance for average burst length of 1 (20 ms), 5 (100 ms) and 25 (500 ms), respectively.

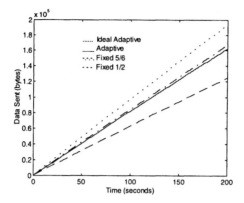

Fig. 4. Comparison of TCP performance for average burst length at 1.

Fig. 4 shows the TCP performance for four coding approaches when the average burst length is one. The result indicates a performance gap between adaptive channel coding based on channel estimation and ideal adaptive channel coding. Even the fixed rate 5/6 code provides slightly better system performance than the adaptive channel coding based on channel estimation. That is due to the delay of coding adaptation to channel condition from channel estimation. When the average burst length is just one data frame and the next frame will most likely have low BER, it is better served by the fixed rate 5/6 code than the adaptive scheme based on history.

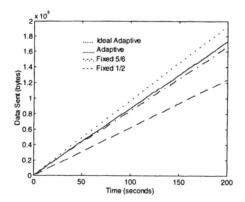

Fig. 5. Comparison of TCP performance for average burst length at 5.

Fig. 5 shows TCP performance for average burst length of 5. The result indicates that adaptive channel coding based on channel estimation provides slightly better system performance than the fixed rate 5/6 code. The performance gap between adaptive channel coding based on channel estimation and ideal adaptive channel coding in such radio environment is smaller than that from average burst length of one as shown in Fig. 4.

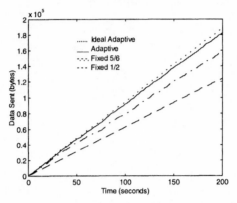

Fig. 6. Comparison of TCP performance for average burst length at 25.

Fig. 6 shows TCP performance for average burst length of 25. The result indicates that TCP throughput performance of adaptive channel coding based on channel estimation is very close to that of ideal adaptive channel coding. It provides significant higher data rate (913 bytes/sec) than a fixed rate 5/6 code (about 795 bytes/sec). The simulation results also indicate that timeout retransmission is very rare for wireless TCP with RLP. For all three values of average burst length, the fixed rate 1/2 code does not provide comparable performance as the others for the studied fading channel. We conclude that adaptive channel coding can effectively improve the TCP throughput in slow fading radio environment (the average burst length more than 5 data frames) while maintaining comparable performance as the fixed rate 5/6 coding for fast fading environment.

5 Conclusions

We have studied performance of circuit-switch based TCP over wireless link. An adaptive RLP is proposed to maintain TCP performance under variable channel condition. Our simulation results indicate significant improvement in throughput performance for TCP over adaptive RLP. The proposed scheme does not require modification of congestion control mechanism in the existing TCP protocol. The main contribution of this study is the consideration of a realistic channel environment and channel quality feedback for the update of RLP coding scheme in determining the throughput of TCP over adaptive RLP. The result is applicable to existing wireless systems where adaptive channel coding can be implemented. The proposed adaptive scheme ensures maximum compatibility with existing systems.

6 References

[1] D. E. Comer, Internetworking with TCP/IP, Vol. 1: Principle, Protocols, and Architecture, Prentice Hall, 1995.

[2] TIA IS-130-A, TDMA Wireless Systems – Radio Interface – Radio Link Protocol 1, July 4, 1997.

[3] H. Balakrishana, V. Padmanabhan, S. Seshan and R. Katz, "A Comparison of Mechanisms for Improving TCP Performance over Wireless Links," IEEE/ACM Trans. Networking, Dec. 1997, pp.756-769.

[4] S. Lin and D. J. Costello, Jr., Error Control Coding: Fundamentals and Applications, Prentice Hall, 1983.

[5] S. Falahati and A. Svensson, "Hybrid Type-II ARQ Schemes for Rayleigh Fading Channels," in Proc. International Conference on Telecommunications, Greece, June 1998, pp. 39-44.

[6] J. Hagenauer, "Rate-Compatible Punctured Convolutional Codes (RCPC Codes) and their Applications," IEEE Trans. Commun., vol. COM-36, Apr. 1988, pp. 389-399.

[7] D. Huang and J. Shi, "Performance of TCP over Radio Link with Adaptive Channel Coding and ARQ," in Proc. 50th Vehicular Technology Conference, Houston, May 1999.

[8] S. Yajnik, J. Sienicki and P. Agrawal, "Adaptive Coding for Packetized Data in Wireless Networks," in Proc. PIMRC'95, pp. 338-342.

[9] V. Jacobson, "Congestion Avoidance and Control," in Proc. ACM SIGCOMM 88, Aug. 1988.

[10] P. Karn and C. Partridge, "Improving Round-Trip Estimates in Reliable Transport Protocols," in Proc. ACM SIGCOMM 87, Aug. 1987.

REDUCING LOCATION UPDATE AND PAGING COST IN A PCS NETWORK

Pablo Garcia Escalle, Vicente Casares Giner, Jorge Mataix Oltra

Departamento de Comunicaciones, UPV.

Camino de Vera, s/n. 46071 Valencia (Spain).

E-mail: pgarciae,vcasares,jmataix[@dcom.upv.es]

Abstract

Mobility tracking operations in personal communication systems are signalling consuming. Several strategies have been proposed in the literature to reduce both, the location update and the paging cost. In this paper we propose a location tracking algorithm called Three Location Area ($TrLA$), combined with selective paging. In the $TrLA$, the mobile terminal (MT) allocates in its own local memory the identification of three LAs. Each LA is composed by a cluster of several cells. Each time the MT enters into a new LA, it triggers a location update message towards the fixed network (FN) and updates its cache memory. Two steps selective paging is also considered and compared with the single step or non selective paging. An analytical model based in a semimarkov process has been used to evaluate our proposal. This new scheme, can easily be implemented in existing standard cellular and personal communication systems, we believe.

1 Introduction

In *Personal Communication Systems* (*PCS*s), by *Mobility Tracking* it is understood as the set of procedures to locate a *Mobile Terminal* (*MT*). An incoming call to a *MT* must be delivered on time at a minimum cost, so the cell where the *MT* is actually roamming must be identified in a short period of time, while keeping the cost under certain constraints. Those procedures are, in *GSM* terminology, *Location Updating* (*LU*), and incoming *Call Delivery*. *Call Delivery* is splitted into *Interrogation* and *Paging* (*PG*) procedures.

Roughly speaking, two *LU* strategies are envisaged in the literature. The static or global strategy, usually implemented in current cellular systems, in which the whole coverage area is divided into several fixed *Location Areas* (*LAs*). The terms *Registration Area* and *Location Registration* are used on the IS-41 standard, instead of *Location Area* and *Location Updating*. Several cells conform a *LA*. Each time a *MT* crosses the borders between two *LAs*, it triggers a *LU* message towards the *Fixed Network* (*FN*) notifying the new visited *LA*. The last recent contact with the *FN* is implemented at the level of *LA*. On the other hand by dynamic or local strategies ([6], [7] and [10]) the last contact with the FN is personalized for each MT at the cell level. In opposite view to the static-global strategy, in the dynamic-local case no division into several LAs is done. In [6], three dynamic-local strategies were proposed. Based on the time elapsed, the number of movements performed or the distance travelled since the last contact with the FN, the MT triggers a LU message.

The *Interrogation* procedure, is supported by the FN. For each incoming call to the MT, the network queries the System Data Base (SDB) of the PCS. The output of this query is the area (a LA, a *cell*, a *Paging Area* (PA), etc..) where the last contact of the MT with the FN occurred. In this paper no attention is payed to the *Interrogation* procedure. After a successful *Interrogation*, follows the PG procedure, by which the MT is paged according to some selective or non selective paging algorithm. The output of the PG procedure is the *cell* where the MT is located in.

In this paper, we deal with *Mobility Tracking* procedures with direct impact in the common air interface (CAI), in particular with LU and with PG procedures. Some work related with the signalling load in FN associated with LU, *Interrogation* and PG procedures, and their impact in the SDB can be found in [8], [9] and [12]. For LU we use fixed LA with hysteresis effect. This avoids "ping-pong" effect when the MT is roamming in the surrounding of a LA border. The multi-layer LU scenario already proposed in [3] is the starting point for our proposal. For PG algorithm we consider *selective* or *multi-step paging* ([5], [11], [13] and [14]). Here we have followed a two steps PG procedure, i.e. the first scenario of [5] with fixed size and static LAs.

The work is organized as follows. In section 2 we present the scenario; the LU and PG procedures are described. In section 3, the mathematical analysis is carried out. Some illustrative examples and discussion are reported in section 4. Finally, in section 5 we address the conclusions.

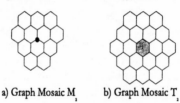

a) Graph Mosaic M$_1$ b) Graph Mosaic T$_2$

Figure 1: Mosaic graphs

		ring 0	ring 1	...	ring m	...
Mosaic T_m	L_m	1	6	...	$6m$	...
	Accumulate	1	7	...	$3m^2 + 3m + 1$	...
	Perimeter	6	18	...	$12m + 6$	...
Mosaic M_m	L_m	3	9	...	$6m + 3$	...
	Accumulate	3	12	...	$3m^2 + 6m + 3$	...
	Perimeter	12	24	...	$12m + 12$	...

Table 1: Number of cells in ring m (L_m), accumulated values and perimeter for different hexagonal cell layout configurations

2 Scenario and procedure descriptions

Since mesh cell layout reports similar values to hexagonal cell layout, as it was proved in [14], for illustrative purposes we have chosen hexagonal cell layout configuration. All cells have the same size, being R the cell radius. The cell residence time of a MT has been characterized with a generalized gamma distribution, [17], with probability density function (pdf) given by

$$f_{cell}(t; a, b, c) = \frac{c}{b^{ac}\Gamma(a)} t^{ac-1} e^{-(t/b)^c}; t, a, b, c > 0 \tag{1}$$

where $\Gamma(a)$ is the gamma function, defined as $\Gamma(a) = \int_0^\infty x^{a-1} e^{-x} dx$ for any real and positive number a. We denote its Laplace Transform (LT) as $f_{cell}^*(s)$ and mean value $-f_{cell}^{*'}(0) = b\Gamma(a + 1/c)/\Gamma(a) = 1/\lambda_m$. After leaving cell i, the MT enters into one of its six neighbouring cells with probability $1/6$. This is in fact a $2D$ random walk mobility model used in several studies.

As in current cellular networks, the whole coverage area is partitioned into a number of fixed (statics) LAs. LAs are configurated with mosaic graphs. Mosaic graphs T and M have been considered as hexagonal configuration [4]. Mosaic graphs are constructed by arranging cells in concentric cycles around a starting point or a cell. When the center is a point or vertex, figure 1-a, we get the so called mosaic graphs M_m where m denotes the outer ring number; the first ring is numbered as 0. When, instead of a point, the center is a cell, figure 1-b, we get dual mosaic graphs, T_m. For T_m, $m = 0$ corresponds to a single hexagon and $m = 1$ corresponds to a cluster of 7 cells. For both mosaic graphs, table 1 shows the length L_m and the accumulated values.

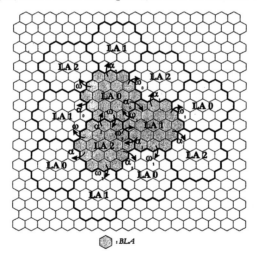

Figure 2: Three LAs each one characterized with a mosaic graph T_2, with their exit probabilities towards neighbouring LAs

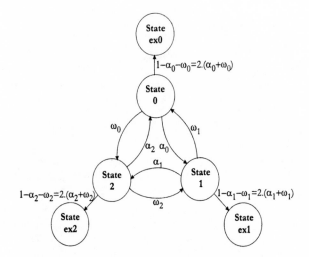

Figure 3: State transition diagram of three LAs

2.1 Location updating procedure

In current cellular networks, each MT keeps in its local memory the record of the *Location Area Identifier* (LAI) where currently is roamming. Each time a MT visits a new LA, it will perform a LU procedure and will update the record of the LAI. One of the main drawbacks of this option is the "ping-pong" effect produced by some MTs with some type of random walk mobility. They can generate too many LU messages when they are roamming in the surrounding of the LA borders. To mitigate this effect, we have adopted the same philosophy as in [3]. It is implemented as follows. The MT will keep in its local memory the identification of three LAs, with some common perimeter between them, see figure 2. This is what we call a *Big Location Area*, BLA. Each time a MT visits a LA it will check if the LAI is coincident with LAI_A, with LAI_B or with LAI_C. If this is the case, no LU message is triggered by the MT. Otherwise, the MT will send a LU message to the FN, containing the new LAI. The SDB will verify the three records of the three LAs of that MT and will order to overwrite in the local memory of the MT one or two old records, whatever corresponds.

2.2 Paging procedure

After each successful *Interrogation* procedure, the system initiates the paging process. When single step is used, the total number of cells, N_c, ($N_c = N_0 + N_1 + N_2$) of the three LAs are paged simultaneously, and a minimum delay in the PG process is achieved. The second alternative is the *Multistep* or *Selective Paging*. In this work we have considered a two steps PG procedure [5]. The MT will first be paged in the LA corresponding to the last recent interaction with the FN. If the

MT doesn't answer, after a certain time out, it will be paged in the other two LAs.

3 Mathematical model

In this section we provide the mathematical analysis to obtain the total cost per call arrival. In section (3.1), the LU cost is derived, and in section (2.2) the PG cost is derived by assuming both, single and two steps paging.

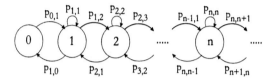

Figure 4: State transition diagram of $2D$ random mobility model used with hexagonal configurations

3.1 Location updating cost

In a first step, we evaluate the *pdf* of the sojourn time in a LA. To that end, we use the state transition diagram shown in figure 4 with the transition probabilities of table 2. Then we evaluate the LU cost using the results of [18], Appendix A. The number of LU messages triggered by the MT between two call arrivals, conditioned to the fact that the MT received its last call in LA_k, $k = 0, 1, 2$, is $Pr(mLU/k)$, [18], Appendix A. Therefore, the expected LU cost per call arrival is given by,

$$C_{up-mt}(k) = P_U \sum_{m=1}^{\infty} mPr(mLU/k) = P_U \mathbf{Pr}_{k,ex}(\lambda_c)\mathbf{T}[\mathbf{I} - \mathbf{ST}]^{-1}\mathbf{e}; k = 0, 1, 2 \qquad (2)$$

where P_U is the LU cost per message triggered by the MT.

	$p_{0,1}$	$p_{1,0}$	$p_{1,2}$	$p_{2,1}$	$p_{2,3}$	...	$p_{n,n-1}$	$p_{n,n+1}$	...
Mosaic T_m	1	$\frac{1}{6}$	$\frac{1}{2}$	$\frac{1}{4}$	$\frac{5}{12}$	...	$\frac{2n-1}{6n}$	$\frac{2n+1}{6n}$	...
Mosaic M_m	$\frac{2}{3}$	$\frac{2}{9}$	$\frac{4}{9}$	$\frac{4}{15}$	$\frac{2}{5}$	...	$\frac{2n}{6n+3}$	$\frac{2(n+1)}{6n+3}$	...

Table 2: Transition probabilities from ring n to ring m $(p_{n,m})$ hexagonal cell layout configurations

Assuming that incoming calls to a MT arrive according to a Poisson process, an arbitrary incoming call will find the MT roaming in LA_k with probability given by,

$$Pr(\text{roaming in } k) = \frac{\frac{1}{\lambda_{m,LAk}}\pi_k}{\frac{1}{\lambda_{m,LA_0}}\pi_0 + \frac{1}{\lambda_{m,LA_1}}\pi_1 + \frac{1}{\lambda_{m,LA_2}}\pi_2} \qquad (3)$$

Expression (3) is derived having into account the PASTA property [2]; π_k and $1/\lambda_{m,LA_k}$ are the portion of visits and the mean sojourn time in state k respectively. Therefore, we uncondition (2) with respect to (3), i.e.

$$C_{up-mt} = \sum_{k=0}^{2} C_{up-mt}(k) Pr(\text{roaming in } k) \tag{4}$$

3.2 Terminal paging cost

As in section 3.1, we will use the results of [18], Appendix A. Between two consecutive incoming calls the MT will trigger (or not) a LU message towards the FN. The terminal PG cost is obtained by considering those two excluding events. We assume the MT received the last incoming call when it was roamming in LA_k.

When two steps paging is applied, the terminal PG cost can be formulated as,

$$\begin{aligned} C_{pg-mt}(k) &= P_V[N_k Pr(\text{no } LU, k/k) + N_c(1 - Pr(\text{no } LU, k/k)) + \\ &\quad \sum_{i=0}^{2} \{N_i Pr(\text{last } LU = i, i/k) + N_c(1 - Pr(\text{last } LU = i, i/k))\}] \end{aligned} \tag{5}$$

where P_V is the terminal PG cost per cell, and N_i is the total number of cells that configurates LA_i. Clearly $N_c = N_0 + N_1 + N_2$. As before, we uncondition expression (5) with respect to (3), i.e.

$$C_{pg-mt} = \sum_{k=0}^{2} C_{pg-mt}(k) Pr(\text{roaming in } k) \tag{6}$$

Obviously, when single step paging (no selective) algorithm is applied, the terminal PG cost can be written as

$$C_{pg-mt}(k) = C_{pg-mt} = P_V N_c \tag{7}$$

Further details can be read in our internal report.

4 Some examples and discussion

The analytical model presented in previous section allows us to obtain the total cost per call arrival C_T under various parameters. It is defined as:

$$C_T = C_{up-mt} + C_{pg-mt} \tag{8}$$

Those parameters include the total number of cells that conforms the three LA, N_c, the LU cost,

P_U, the PG cost, P_V, the call arrival rate λ_c, the mean cell residence time $1/\lambda_m$ and therefore the call-to-mobility ratio defined as $CMR = \lambda_c/\lambda_m$. The residence time for a cell has been assumed to be characterized by a generalized gamma distribution. To that end we have followed the approach proposed in [17].

In figures 5, 6, 7 and 8 we report the total cost per call arrival, C_T, for some $CMR = 0.01, 0.1, 1$ and 10. As in [14], we have chosen $P_U = 10$, $P_V = 1$. All LAs have the same size. Mosaic graph M and T have been considered (LA configuration). For low CMR, let us say $CMR < 1$, we observe an optimum value for the LA size, N_c^*. As N_c exceeds its optimum value, N_c^*, the terminal PG cost dominates and C_T is an increasing function of N_c. And viceversa, as N_c decreases from its optimum value, the LU cost dominates. When CMR becomes higher than 1 the optimum LA size is reduced to 1, because of the low mobility of the MT. For a fixed value of CMR a significant saving is also achieved when the paging delay increases from 1 to 2.

For low CMR, $CMR = 0.01$ and single step paging, delay 1, the optimum configuration is $1LA/BLA$, mosaic graph T, $N_c = 61$. The total cost is approximately $C_T = C_{up-mt} + C_{pg-mt} \cong 147 + 61 = 208$, figure 5. It is worth to note that $2LA/BLA$ and $3LA/BLA$ configurations are worst than the $1LA/BLA$ configuration. It can be explained, for instance, if we compare a) $1LA/BLA$, mosaic graph M, $N_c = 75$, with b) $2LA/BLA$ mosaic graph T, $N_c = 2 \times 37 = 74$. Both configurations roughly present the same total number of cells, therefore roughly the same PG cost. However the LU cost in case b) is higher than LU cost in case a), due to the fact that the perimeters are, 60 edges for $1LA/BLA$ and 70 edges for $2LA/BLA$, see table 1.

When two steps paging, delay 2, is considered, the conclusions are quite different. According to the TLA strategy presented in [16], $2LA/BLA$, mosaic graph T, $N_c = 2 \times 61 = 122$, shows the optimum total cost, equals to $C_T = C_{up-mt} + C_{pg-mt} \cong 123 + 73 = 196$. A better result is obtained for $3LA/BLA$ scenario, mosaic graph T, $N_c = 3 \times 37 = 111$, where the optimum total cost, equals to $C_T = C_{up-mt} + C_{pg-mt} \cong 126 + 57 = 183$. When comparing with the optimum value of $1LA/BLA$, single step, we achieve a saving total cost of $(208 - 196)/208 \cong 0.0577$, (5.77%), and $(208 - 183)/208 \cong 0.12$, (12.02%), respectively.

5 Conclusions

This paper deals with *Mobility Tracking* strategies. In the literature they are classified as static or global and dynamic or local. Our new strategy can be seen as an hybrid between the global and local strategies. It can be included in the first research area cited in [15], at the bottom of page 140, i.e. as an option to improve the performance of *location management* strategies of actual real cellular systems, such as IS-54, IS-95 and GSM, to name a few. In fact, we believe that our proposal can be easily implemented in those real cellular systems.

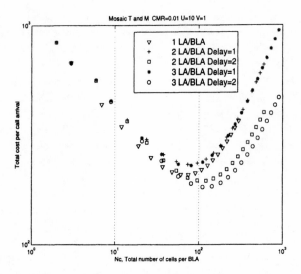

Figure 5: Total cost (location +paging) per call arrival. $CMR = 0.01$, $P_U = 10$ and $P_V = 1$.

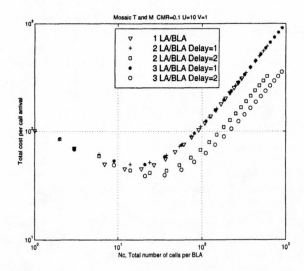

Figure 6: Total cost per call arrival. $CMR = 0.1$, $P_U = 10$ and $P_V = 1$.

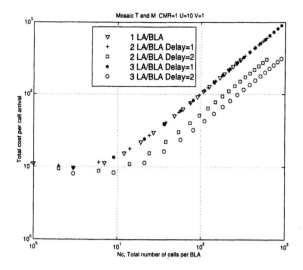

Figure 7: Total cost per call arrival. $CMR = 1$, $P_U = 10$ and $P_V = 1$.

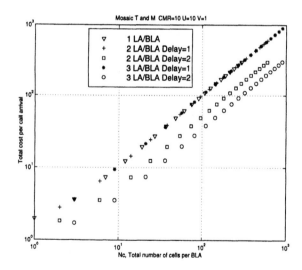

Figure 8: Total cost per call arrival. $CMR = 10$, $P_U = 10$ and $P_V = 1$.

6 Acknowledgments

This work has been financed by *Comisión Interministerial de Ciencia y Tecnología Spain) (CICYT)* under project $TIC98 - 0495 - C02 - 02$. The authors thank $CICYT$ for financial support in the presentation of the paper.

References

[1] L. Kleinrock "Queueing Systems". Vol. 1 John Wiley, 1975.

[2] R. W. Wolf, "Poisson arrival see time averages" Operation Research, 1982

[3] S. Okasaka, S. Onoe, S. Yasuda, A. Maebara "A new location updating method for digital cellular systems", Proceedings of the IEEE 41st Vehicular Technology Conference pp. 345-350. May 19-22, 1991. St. Louis, Missouri (USA).

[4] E. Alonso, K.S. Meier-Hellstern, G. Pollini, "Influence of cell geometry on handover and registration rates in cellular and universal personal telecommunication networks, "Proceedings of the 8th ITC Spec. Sem. Universal Pers. Telecommun. pp. 261- 270, October 12- 14, 992. Santa Margherita Ligure, Genova (Italy).

[5] F. V. Baumann, I. G. Niemegeers "An evaluation of location management procedures", Proceeding of the 1994 third I.C. on Universal Personal Communications, pp.359-364. September 27, October 1, 1994 San Diego, California (USA).

[6] A. Bar-Noy, I. Kessler, Mo. Sidi, "Mobile users: To update or not to update", Proceedings of the INFOCOM'94 pp. 570-576, June 14-16, 1994, Toronto, Ontario (Canada).

[7] A. Bar-Noy, I. Kessler, Mo. Sidi, "Tracking strategies in wireless networks", Proceedings of the IEEE Int. Symp. on Information Theory, pp. 413, June 27- July 1, 1994. Trondheim (Norway).

[8] Y.B. Lin, "Determining the user location for personal communications services networks". IEEE Trans. on Vehicular Tehcnolgy Vol. 43, n. 3 pp. 466 -473. August 1994.

[9] G. P. Pollini, K.S. Meier-Hellstern, D. J. Goodman, "Signalling traffic volume generated by mobile and personal communications". IEEE Communications Magazine, Vol. 33, n.6. pp. 60-65, June 1995.

[10] A. Bar-Noy, I. Kessler, Mo. Sidi, "Mobile users: To update or not to update", ACM-Baltzer Journal of Wireless Networks, Vol. 1, n. 2, pp. 175-186, July 1995.

[11] I. F. Akyildiz, J. S.M. Ho, "Dynamic mobile user location update for wireless PCS networks", ACM-Baltzer Journal of Wireless Networks, Vol. 1, n. 2, pp. 187-196, July 1995.

[12] R. Jain, Y.B. Lin, "An auxiliar user location strategy employing forwarding pointers to reduce network impact of PCS", ACM-Baltzer Journal of Wireless Networks, Vol 1, n. 2, p.197-210, July 1995.

[13] J. S.M. Ho, I. F. Akyildiz, "Mobile user location update and paging delay constraints", ACM-Baltzer Journal of Wireless Networks, Vol. 1, n. 4, pp. 413-425, December 1995.

[14] I. F. Akyildiz, J. S.M. Ho, Y-B Ling. "Movement-Based location update and selective paging for PCS networks", IEEE/ACM Trans. on Networking, Vol. 4, n. 4 pp 629-638, August 1996.

[15] I. F. Akyildiz, J. S.M. Ho. "On location management for personal communications networks" IEEE Communication Magazine, Vol. 34, n. 9 pp. 138-145, September 1996.

[16] Yi-Bing Lin, "Reducing location update cost in a PCS", IEEE/ACM trans. on Networking, Vol 5, n.1 pp.25-33, February 1997.

[17] M. M. Zonoozi, P. Dassanayake. "User mobility modeling and characterization of Mobility patterns", IEEE JSAC Vol, 15, n. 7, pp 1239-1252, September 1997

[18] P. Garcia, V. Casares, J. Mataix. "Reducing Location Update and Paging Cost in a PCS Network", Internal Report, Departamento de Comunicaciones, UPV, March 1999

Performance Enhancement for TCP/IP on Wireless Links[*]

J. Scott Stadler, Jay Gelman, and Jeffery Howard
MIT Lincoln Laboratory
Lexington, MA 02420
stadler@ll.mit.edu

ABSTRACT- *The TCP/IP protocol suite that forms the basis of the Internet was designed to operate over an extremely large range of environments. Despite this robustness, degraded levels of performance may be experienced when the assumptions inherent in its algorithms are violated. For instance, the potentially high BER of a terrestrial wireless link or the high delay-bandwidth product of a satellite link result in a situation in which the link is not efficiently utilized and the TCP/IP performance (as perceived by an interactive user) may be poor.*

This paper will identify the reasons for the reduced levels of performance and describe two techniques for dramatically improving the performance (efficiency and perceived QoS) of TCP/IP in a wireless environment. The selected techniques possess the desirable properties of transparency, scalability, and backward compatibility. The first is a link layer protocol that uses a combination of fragmentation coupled with selective repeat ARQ at the link layer to improve TCP/IP performance. The second is a split connection approach that eliminates TCP from the wireless segment of the network. This allows protocols developed specifically for the wireless environment of interest to be used on the wireless segment of the network path. Both of these techniques have been prototyped and tested on the Wireless Networking Test-bed at Lincoln Laboratory. Test results show that nearly optimal performance may be obtained for many scenarios of interest.

1 INTRODUCTION

The extension of TCP/IP networks via wireless has a number of important applications for personal, business, and military users. Wireless networks can be used to service mobile users via a fixed terrestrial infrastructure (PCS) or to provide an instant communications infrastructure almost anywhere in the world via satellite. The later is especially important when it is necessary to connect rural areas, whose population density would not warrant the construction of a large wired plant, into the global network.

While the use of wireless provides a very flexible way to extend networks, there are a number of technical issues that need to be addressed. These issues revolve around the fact that most protocols were optimized to run on a wired (fiber/copper) infrastructure. The goal is therefore to develop protocol enhancement techniques that allow standard protocols to be seamlessly extended over wireless links. One such technique, the Wireless IP Suite Enhancer (WISE) which is a transparent protocol enhancer was introduced in [1]. Test results showed that WISE offers near optimal performance on the processed military satellite links for which it was developed [2]. The focus of this paper will be the experimental evaluation of WISE in commercial satellite and terrestrial wireless environments.

The paper begins with a brief description of the problems encountered when a TCP/IP network is extended over a wireless link. This leads to a detailed description of the Wireless IP Suite Enhancer, which was designed to transparently overcome these problems. The predicted performance gains are then experimentally verified on a Wireless Networking Tested. The test bed is capable of evaluating the performance of various protocols in a simulated (HW and SW) environment. It currently supports non-

---

[*] This work was sponsored by the Department of the Air Force under contract number F19628-95-C-0002. Opinions, interpretations, conclusions, and recommendations are those of the author and are not necessarily endorsed by the United States Air Force.

enhanced TCP both with and without elective protocols, a link layer enhancer [8], and WISE. Finally, experimental results for commercial satellite and terrestrial wireless systems are presented showing that WISE yields nearly optimal performance in those scenarios. Unlike many previous studies that focus on "best case" performance for specially tuned TCP implementations, this paper focuses on "typical case" performance in realistic operational scenarios.

2   TCP via Wireless

The TCP/IP protocol suite that forms the basis of the Internet was designed to operate over an extremely large range of environments. Despite this robustness, degraded levels of performance may be experienced when the assumptions inherent in its algorithms are violated. The reduced performance can be primarily attributed to three characteristics of wireless links; bit errors, latency, and asymmetry. A cursory treatment of these effects will be given here with a detailed treatment appearing in [1]. A thorough exposition of the intricacies of the TCP/IP protocol suite can be found in [3].

Bit Errors - Most networking protocols were designed for use in a terrestrial environment were the BER is extremely low (typically less than 10^{-10}). On a wireless link, the raw BER is much higher typically ranging from 10^{-2} to 10^{-6} and can result in very poor TCP link utilization.

Latency - Latency is primarily an issue for satellite links. Latency in a terrestrial environment is typically very low (prorogation time across the US is about 30 ms). The latency through a geo-synchronous satellite hop (up and down) is about 260 ms resulting in a Round Trip Time (RTT) of about 520 ms plus coding delays and any additional terrestrial based latencies. The effects of a large RTT on TCP's feedback mechanisms is a primary impediment to efficient link use on satellite networks.

Asymmetry - Unlike terrestrial networks, wireless links often operate in an asymmetric mode, receiving a higher data rate than they transmit. This is due to the fact that wireless terminals (especially portable/mobile) are often uplink power limited. Link asymmetry will interact with TCP's self-clocking mechanism. This is because most TCP segments sent on the high rate link result in an ACK being sent on the low rate link. The result is that the low rate link may become congested with ACKs causing a reduction in the amount of data being sent on the high rate link even if the high rate link is not congested.

Several TCP elective protocols reduce the impact of the above mentioned impairments. These modifications are detailed in "TCP Extensions for High Performance" [4], "TCP Slow Start, Congestion Avoidance, Fast Retransmit, and Fast Recovery Algorithms" [5], and "TCP Selective Acknowledgement Options" [6]. These algorithms will improve the performance of TCP connections on a wireless link in many instances, but they are ELECTIVE. This means that a given TCP implementation may choose not to implement some or all of the protocols. If **both** hosts participating in a TCP connection do not implement the protocols, then neither will experience an improvement in performance. Furthermore if the applications (i.e., client/server) are not developed with these options in mind, then they may not be able to take advantage of them (e.g., An application may override a large default window size with a small window size).

3   Goals

There are a number of goals that must be considered in the evaluation of alternate methods of improving TCP/IP performance on wireless links. A brief description pertaining to the issues associated with each goal is given below.

Transparency – By transparency, we mean that the existence of a protocol enhancer/booster will have no impact on the connections that pass through the enhancer with the exception that the performance will be improved. In addition, the end user should not need to have any knowledge that the enhancer/booster is being used nor should the end user need to follow special procedures to obtain performance improvements.

Backward Compatibility – The protocol enhancer must work with the existing Internet and intranet infrastructures. It is unrealistic to expect the entire Internet community to change/upgrade their protocol stacks and or applications to accommodate users who access the network via wireless.

Efficiency – While wireless links have many advantages over wired links with respect to their ability to provide mobile connectivity and instant infrastructure, they are typically more expensive than wired alternatives. Because of this it is important that the wireless link be used in as efficient a manner as possible.

Scalability – The enhancing approach should scale to high data rates and large numbers of users. While approaches with limited scalability may satisfy near term needs, they can quickly become obsolete in a market that has been experiencing exponential growth.

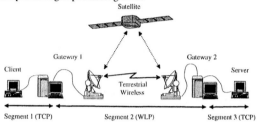

Figure 1 – WISE System Configuration

4 Wireless IP Suite Enhancer

The Wireless IP Suite Enhancer (WISE) improves the performance of the TCP/IP protocol suite in a wireless environment, increasing the wireless link utilization and dramatically improving the performance. The WISE approach consists of software that is added to gateways at the periphery of the wireless segment of the network, Figure 1. In a satellite-based system, these gateways are separate entities from the server or the client as a single gateway typically services an entire WAN. For personal mobile systems however, the server side gateway would play the role of a foreign agent and WISE would operate concurrently with the traditional mobility management functions. The client side gateway functionality, however, would be physically co-located with the wireless host (i.e., The client and client side gateway functionality would both be implemented in a single PCS device.).

WISE operates by transparently splitting the TCP connection into three segments, client to gateway, gateway to gateway via the wireless link, and gateway to server, Figure 1. The client to gateway and gateway to server segments use terrestrial connections and operate using unmodified TCP/IP protocols. The gateway to gateway connection, however, uses a special Wireless Link Protocol (WLP) developed according to the physical characteristics of the wireless link at hand. The WISE software is responsible for converting TCP to WLP upon entering the wireless sub-network and back to TCP upon exiting.

There are many advantages to converting TCP to another protocol at the WISE gateway. Since there is no concept of TCP on the high delay and/or error rate segment of the network, the detrimental effects of latency and errors on TCP are avoided and link utilization is greatly increased. The TCP fairness problem is also circumvented, as TCP connections do not span the satellite. The TCP/IP headers are replaced with a much shorter WLP header and compression of the TCP/IP data may be performed so that fewer bytes need to be sent over the wireless segment. In addition, encryption can be used to protect the data from eavesdropping. Finally, the system can be implemented without making any changes to the TCP/IP code on the gateway, and NO changes of any kind are required to the end users protocols or applications.

The WISE system is very flexible and may be configured in several different manners depending on the network topology and the type of encryption that will be used. In a transparent configuration, it may be inserted anywhere within an IP network (subject to some weak topological constraints) and will boost performance without making any changes outside of the wireless segment. It can also be configured in a "last hop" mode for individual wireless hosts as described above in which case WISE must reside on the wireless host. WISE may also be used in systems with asymmetric and/or mixed media links such as those described in [7].

WISE has two components, the connection splitting mechanism and the Wireless Link Protocol (WLP). The Lincoln Laboratory Link Layer (LLLL) was chosen as the WLP. It provides a reliable connection oriented service to the higher layers, which matches well with the service provided by TCP [8]. Both the WLP and the connection splitting mechanism are detailed below.

4.1 Wireless Link Protocol

The Lincoln Laboratory Link Layer provides reliable and ordered connection oriented delivery of packets across a wireless link. It also incorporates features that enhance TCP performance and efficiency. It is important to note that the LLLL will enhance TCP performance when used outside the scope of WISE. This is necessary, as the WISE system will not be able to split some connections (i.e., Any connection for which the TCP header cannot be read). While the performance of these connections will not improve to the same extent as connections that are split, they will still be enhanced by the LLLL.

The approach taken by the LLLL is threefold. When a new connection is created get data flowing as quickly as possible. Once data starts flowing prevent TCP's flow control algorithm from reducing the flow. This is necessary because the flow control algorithm will confuse errors with congestion and reduce flow when no congestion exists. Note that when congestion is present on either the terrestrial or satellite portions of the link, flow will be reduced, as it should be. Finally, errors must be corrected in as efficient a manner as possible.

The guidelines outlined in the previous paragraph are accomplished using a combination of fragmentation and selective repeat ARQ both of which are performed at the link layer. The use of fragmentation decouples the TCP segment size from the link layer packet size. This is important because the larger the TCP segment size, the quicker the flow control algorithm will inject data into the network. On the other hand, larger packets are more susceptible to errors and result in larger amounts of data being retransmitted when an error does occur. Fragmentation allows large TCP segments to be used, while at the same time retaining the benefits of smaller link layer packets. Note that the fragmentation is restricted to the wireless segment of the network and that it is done transparently to IP (similar to the fragmentation of IP packets into ATM cells for IP/ATM).

Selective Repeat ARQ is used to hide any link errors from TCP and thus prevents TCP from miss-interpreting errors as congestion. Link layer acknowledgments containing the entire state of the receive buffer are periodically sent from the receiver to the sender. Because the ACKs are interrupt driven as opposed to data driven, the link layer will work well on asymmetric links (Note that TCP ACKs are still data driven so TCP may still experience asymmetry problems when the connection is not split.). When a packet does need to be retransmitted, it is retransmitted several times (typically 2-3) to insure that the packet will not need more than a single retransmission (This is a very simple form of FEC.). This is important as repeat transmissions will typically cause TCP's timers to expire, negating any benefit of link layer retransmissions. Selective repeat ARQ results in packets being received out of order, therefore packet reordering is performed prior to defragmentation by the receiving link layer process.

The use of the LLLL in WISE is two-fold. For connections that can be split, it simply provides a reliable connection from one WISE node to its peer node at the other end of the satellite. For connections that cannot be split, it provides an enhanced link that aids TCP by shielding it from link errors.

4.2 Connection splitter

The splitting of TCP connections is illustrated in Figure 2. Here, one or more remote users may communicate with each other or with the terrestrial network via a wireless link. The protocol stack configurations are indicated along the bottom. Note that both of the end users have standard commercial TCP/IP protocol stacks and will run without modification. In addition, they need no knowledge that they are communicating through a wireless link and will not need to use any special procedures. At the periphery of the wireless portion of the network, gateways will be used. Since these gateways will be small in number and they will be aware of the presence of the wireless link, the burden encountered in configuring them will be small and within the scope of the organization providing the "TCP/IP via Wireless" service.

The modified gateways will perform the protocol translation from TCP to WLP for incoming packets and WLP to TCP for outgoing packets. The WLP is responsible for providing a reliable connection oriented link between gateways. Error control may be accomplished using either forward error control coding or with the use of ARQ as outlined above. Note that because WISE performs true protocol conversion, there are no TCP or IP headers to be transmitted on the wireless link, reducing the data rate required.

The system operation is as follows. IP packets not containing TCP segments (or whose TCP headers cannot be read) that arrive at the periphery of the wireless network will go up the protocol stack to the IP layer where the standard routing functions will be performed. The packet will go down the protocol stack through the LLLL and over the wireless link. Any errors encountered during transmission will be corrected by the two peer LLLL layers transparently to IP. TCP packets on the other hand, will pass all the way up the protocol stack to the WISE server. Here the TCP connection will be terminated and a virtual circuit will be set up through the LLLL to the WISE server on the other side of the wireless link. This will cause the receiving WISE server to establish a TCP connection to the intended recipient. Once the connections are all established the data is passed over the wireless link to the receiving WISE server where it is relayed (via the TCP connection) to the intended recipient.

WISE has a number of advantages over other approaches. Because the TCP connection is terminated before the wireless link, none of the problems associated with sending TCP via wireless are encountered. TCP and IP headers as well as TCP segments containing only ACKs are not sent on the

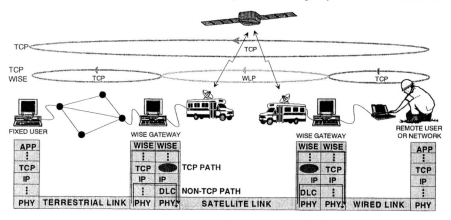

wireless link reducing the needed BW and eliminating problems associated with asymmetry. In addition, WISE can perform arbitrary compression and/or encryption of the TCP data prior to sending it on the wireless link. Most importantly, WISE is completely backward compatible with existing protocols and applications on the user machines (RFCs 1323, 2001, 2018 may be used but are not needed) and requires no modification to the TCP/IP code on the gateway machines.

Encryption techniques that hide the TCP header may have an impact on WISE operation. WISE is fully compatible with application layer encryption and Secure Socket Layer (SSL) encryption, but when operating in a transparent mode, WISE relies on an ability to gather information from the TCP header. If this information is encrypted, then WISE will not be able to split the connection, and only WLP performance gains will be experienced. IPSEC is one such instance where transparent splitting is not generally possible. IPSEC can, however, be handled if the gateways can be considered trusted hosts as would be the case for private network extension.

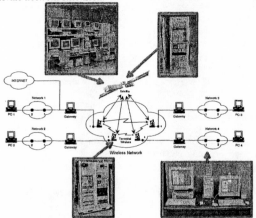

Figure 3 - Wireless Networking Test-bed

5   Wireless Networking Test-bed

A Wireless Networking Test-bed has been developed by Lincoln Laboratory to test the performance of various protocols and architectures in realistic satellite and terrestrial wireless environments. The test-bed is depicted in Figure 3. It consists of a wired network and several remote wireless networks. The topology can be configured as needed to support any testing scenario of interest. The wireless segment is a hardware emulation (at RF) of a processing satellite with an onboard access controller that dynamically allocates resources on demand. It also has an onboard packet switch that is capable of switching packets from various uplinks to the appropriate downlinks. Note that while the satellite emulator offers several advanced capabilities, it can also be configured to look like a standard transponded satellite or a simple terrestrial wireless link. The test terminal can be configured to emulate one or more standard terminal types and includes FEC capabilities. The gateway is responsible for interfacing a terrestrial network consisting of one or more unmodified hosts to the wireless segment of the network. It also implements the protocol enhancer. Finally, the user host is an unmodified personal computer running commercial protocols and applications. The test-bed is instrumented to allow the characterization of system performance at any layer in the protocol stack from physical to application and at any point in the network. The test-bed has been used to successfully demonstrate WISE with various applications such as Telnet, FTP, and WWW browsers.

In addition to a hardware emulation of the wireless link, a software channel simulator (SCS) was also developed. The SCS is a generic UNIX System V *STREAMS* module that emulates the limited data rate, long delay (for satellite channels) and higher error rates of wireless channels. The error burst statistics may be set to accurately emulate the distribution of errors that occur due to fading and/or the use of FEC. The SCS can be used with any protocol and its parameters are tunable over a wide range to match various scenarios of interest.

5.1 Automated Test System

The Wireless Networking Test-bed is capable of emulating any reasonable scenario of interest, but the actual test execution needed to compare various approaches for a given scenario can be tedious. The Automated Test System (ATS) is a suite of client-server applications that automate data collection, create test databases, extract results from the databases, and plot the results. The ATS operates from a configuration file that describes the desired ranges for the dependent variable(s) in a given test. From this file, the ATS will automatically perform the following functions: 1) configure the Wireless Networking Test-bed scenario, 2) configure the protocol stacks on the hosts beings tested, 3) execute the test runs (including background loading and data collection), and 4) store the data. The results are stored in a database that can be accessed offline by a separate extraction utility. Using a description of the desired independent and dependent variables, this utility will search through the database and pull out all of the pertinent data and then create a Matlab script to automatically create a plot.

The ATS is capable of controlling the following dependent variables: BER, latency, data transfer rate, file size, application protocol type (e.g., HTTP, FTP), the number of simultaneous transfers (e.g., cross traffic), and the number of trials to be performed. It will support the following protocol stacks: 1) Unmodified TCP, 2) TCP with the SACK/LFN extensions, 3) TCP with LLLL enhancement [8], and 4) TCP with WISE enhancements. The physical layer parameters (e.g., BER, latency, and data rate) may be independently set for the forward and return paths allowing asymmetric and mixed media scenarios to be explored.

The following dependent variables are collected and stored for each transfer:

Transfer Time. The total number of seconds to send a file (including application overhead)
Channel Utilization. The number of bits that flow across the wireless link in a second divided by the data rate of the connection.
Goodput. The number of bits of application data that flow across the wireless link divided by the data rate of the link. Goodput is similar to channel utilization, but takes into account *all* of protocol overhead (TCP, IP, LLLL, and WISE).
TCP Timeouts. The number of TCP timeouts as a function of time during the transferring a file.
Window Size. The evolution of the sender's and the receivers window.
RTO Estimation. The evolution of the sender's and the receiver's timeout which is based on an estimate of the length of one round trip time through the connection.

The automated test system is setup on four workstations as depicted in Figure 4. In this configuration, the SCS is used to emulate the physical layer effects (i.e., link rate, error rate, and latency). The application layer of the survey system is responsible for controlling the configuration of the tests and their execution. The system is divided into three processes across four hosts, server, gate, and client. The tasks performed by each process are described in the following paragraphs.

The client carries the intelligence of the testing suite, including the control logic for the test. It is therefore responsible for directing the server and the gate processes in how to set up and what actions to perform during the course of a test run. The client also collects the statistics that were tracked, storing them in a data-base that can be accessed on a per test or a per run basis.

The server handles configuration of the sending side and implements the application layer server (e.g., HTTP, or FTP). It is also responsible for collection of statistics at the server side such as the sender's window size and RTO estimate. The server process accepts commands from the client just before and after each transfer to start and stop collecting these statistics. Additionally, whenever the client needs to specify a new file size to be transferred, it sends the size to the server, which generates the appropriate sized file.

The gate processes run on the two machines in the center of the Figure. They set the physical characteristics of the (simulated) wireless link and configure the gateways to provide no enhancement, LLLL enhancement, or WISE enhancement. The gateway closest to the client collects statistics regarding the transfers that pass by it.

Although the client must frequently contact the other processes in the test suite, it does not maintain a persistent connection to any of them. Since the protocol stacks must be brought down and reconstructed with each new setup, the system follows a pattern of dominos. First, the client communicates the new file size to the server. Secondly, the farthest gateway from the client hears of the new configuration. It disconnects itself from the network in the process of tearing down its protocol stack and building a new one. The same thing happens to the nearby gateway. When the client is satisfied that enough time has passed for all the servers and gateways to be reconfigured and ready, it will begin the next round of transfers. During that time, in order to avoid interference with the transfer under measurement, no connections are maintained from the client to the gateway or server other than the transfer being recorded.

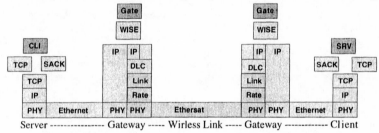

Server ---------------- Gateway ----- Wirless Link ----- Gateway ------------ Client

Figure 4 – Automated Test System Protocol Configuration

6   Performance Results

This section gives a sampling of some of the HTTP test results obtained on the Wireless Networking test-bed under 4 different configurations: 1) TCP with no enhancements (TCP), 2) TCP with only LLLL enhancement (TCP/LLLL), 3) TCP with WISE enhancement (TCP/WISE), and 4) TCP with elective protocols [4-6] but no enhancements (TCP/SACK-LFN). For all tests, the default TCP parameters were accepted (window sizes were 17,520 for TCP and 18,824 for TCP/SACK-LFN). While we are well aware that substantially improved performance could be obtained by tuning TCP parameters, the focus of this paper is on *typical case* performance. This approach is relevant because it represents the performance that can be achieved if the client or server are not accessible for tuning (i.e., controlled by a different party) or are not aware they are using a wireless link (i.e., a satellite in the Internet backbone).

The first set of results are used to parameterize the performance of TCP as a function of the wireless link BER and the round trip time (RTT). For the BER tests, the channel rate was 1 Mbps, the RTT was 0.6 seconds, and the length of the file transfers was 1 Mbyte. Figure 5 shows the average transfer time (over 20 trials) and the average number of TCP timeouts as a function of the BER. It can be seen in the figure that for a BER less than about 2×10^{-6}, that the performance of TCP, TCP/LLLL, and TCP/SACK-LFN are all approximately equal indicating that BER is not significant in this realm. TCP/WISE performance is somewhat better ($\sim$ 12 seconds) due to the fact that it eliminates the effects of delay on the TCP Slow Start algorithm and because TCP/IP headers are not sent over the channel. For BERs greater than 2×10^{-6}, however, the curves start to diverge. At this point, TCP and TCP/SACK-LFN can no longer react appropriately to the bit errors and the performance breaks down. The performance of TCP/LLLL is marginally worse but is still reasonable even down to a BER of 10^{-4}. TCP/WISE is virtually unaffected by the BER over the entire range. In fact the only impact BER has on TCP/WISE operation is that channel goodput is somewhat reduced due to the increase in the number of LLLL fragments that must be retransmitted. Figure 5b yields some insight into the cause of the difficulties experienced by TCP and TCP/SACK-LFN. This plot shows the average number of TCP timeouts experienced during a transfer. For TCP and TCP/SACK-LFN, the number of timeouts starts to increase rapidly as the BER decreases. When TCP is run over the LLLL, however, the number timeouts experienced remains relatively stable. For TCP/WISE which is flow controlled between the client/server and the gateway, no TCP timeouts are observed (i.e., the curve is on the x-axis).

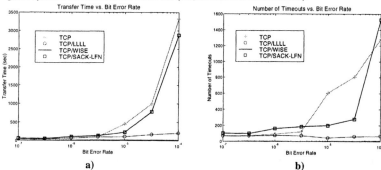

a)   b)

Figure 5 – a)Transfer time vs BER for 1 Mbyte file and a 0.6 second RTT, b) Total number of TCP timeouts exterienced during the transfer.

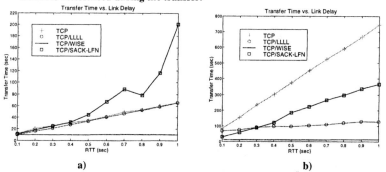

a)   b)

Figure 6 –Transfer time vs RTT for 1 Mbyte file a) BER = 10^{-7} b) BER = 10^{-5}.

The impact of the RTT on protocol perfromance appears in Figure 6 for the same 1 Mbps satellite link. The BER was set to 10^{-7} in Figure 6a and 10^{-5} in Figure 6b. For low BER, the transfer time increases with the RTT for all of the protocols except TCP/WISE which exhibits almost no dependancy on theRTT. This is because for TCP/WISE, RTT only effects the initalization of a transfer and the recovery time for retransmitted packets while for the other protocols the amount of time needed before the entire channel can be utilized is heavily depenent on the RTT. When the BER is increased to 10^{-5} the difference is much more significant for TCP and TCP/SACK-LFN while TCP/LLLL and TCP/WISE are nearly immune to RTT effects as is evidenced by the nearly horizontal lines, Figure 6b.

The next set of results show the effects of file length on the total transfer time at a BER of 10^{-7} and 10^{-5} for the satellite scenario. Once again we see that there is little difference among TCP, TCP/LLLL and TCP/SACK-LFN for a BER of 10^{-7} but that the curves have different slopes and quickly diverge at 10^{-5}. This is due to the fact that TCP/LLLL and TCP/SACK-LFN open their windows wider and spend more of the transfer at higher levels of utilization. WISE on the other hand yields very good perfromance due to the fact that is can instantaneously utilize the entire channel data rate without having to go through a low utilization Slow Start period as is the case for the other approaches.

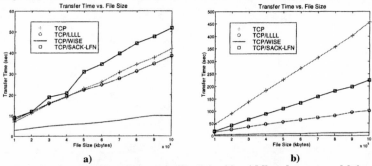

a)   b)

Figure 7 – Transfer time vs file length for a satellite link with a 1 Mbps data rate and 0.6 second RTT a) BER = 10^{-7}, b) BER=10^{-5}.

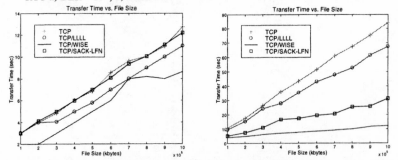

a)   b)

Figure 8 – Transfer time vs file length for a terrestrial wireless link with a 1 Mbps data rate and 0.1 second RTT a) BER = 10^{-7}, b) BER=10^{-5}.

The transfer time vs file length was also explored for the terrestrial wireless case with a 0.1 second RTT. The results for a BER of 10^{-7} and 10^{-5} appear in Figures 8a and 8b respectively. In the low

BER (10^{-7}) tests, all of the protocols performed relatively well. This is because the latency is low enough that all protocols can quickly fill the entire link. Once the BER increases to 10^{-5}, however, TCP and to a somewhat lesser extent TCP/LLLL are repeatedly forced into slow start while TCP/SACK-LFN can quickly react to errors and maintain a large window size.

The instantaneous utilization of the link was also explored under the various scenarios. A utilization plot for a single transfer appears in Figure 9a with the corresponding timeouts appearing in Figure 9b for the terrestrial wireless scenario with a 10^{-5} BER. The utilization starts at zero when the connection is made, increases and levels off during the transfer, and then returns to zero when the transfer is complete. It is clear from the figure that TCP/WISE performance is nearly ideal with an almost instantaneous turn on to 100% utilization followed by a sharp transition back to zero at the end of the transfer. The number of TCP timeouts observed during the transfer shows that the LLLL is reducing the total number of timeouts, but TCP/SACK-LFN can react quickly to the errors and thus provides better overall performance in a low latency environment.

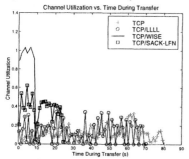 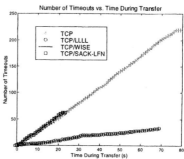

Figure 9 – Perfromance for a terrestrial wireless link with a 1 Mbps data rate and 0.1 second RTT, and a 10^{-5} BER a) Instantaneous utilization, b) TCP timeouts.

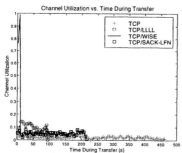 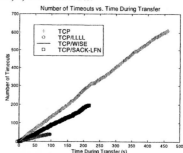

Figure 10 – Perfromance of a 1 Mbyte HTTP transfer over a satellite link with a 1 Mbps data rate, a 0.6 second RTT, and a 10^{-5} BER a) Instantaneous utilization, b) TCP timeouts.

Results for a satellite link with a BER of 10^{-5} and a 0.6 second RTT appear in Figure 10 for a single 1 Mbyte transfer and in Figure 11 for 5 simultaneous 1 Mbyte transfers (Note that these plots were smoothed by averaging over 5 second intervals.). For the single transfer case, WISE makes nearly optimal use of the satellite link and experiences no TCP timeouts. The performance for TCP and

TCP/SACK-LFN is poor. TCP/LLLL on the other hand is somewhat better than TCP and TCP/SACK-LFN due to its ability to eliminate TCP timeouts.

For the case of five simultaneous transfers, the results are similar except that we see somewhat higher levels of utilization from TCP, TCP/LLLL and TCP/SACK-LFN. This is due to the fact that parallel transfers will fill up more of the channel. Once again, TCP experiences the largest number of timeouts followed by TCP/SAC-LFN, and TCP/LLLL with no timeouts for TCP/WISE.

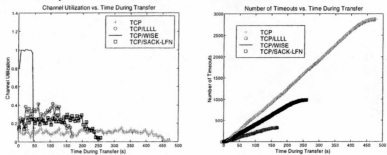

Figure 10 – Perfromance of 5 simultaneous 1 Mbyte HTTP transfers with a 1 Mbps data rate, a 1 second RTT, and a 10^{-5} BER a) Instantaneous utilization, b) TCP timeouts.

7 Conclusions

MIT Lincoln Laboratory has devoted considerable effort to solving the problems associated with extending networks via wireless. Lincoln's research program has resulted in the development and test of the Wireless IP Suite Enhancer that significantly improves performance in a wireless environment by incorporating additional software at the gateways.

Several general observations can be drawn from the data presented: 1) At low BER, the difference among the various approaches is reduced lessening the need for enhancement techniques. 2) At higher BER the TCP elective protocols operate well as long as the RTT is not high. 3) A link layer retransmission strategy is effective in reducing the number of timeouts seen by TCP (and could also be used to aid TCP/SACK-LFN). 4) A connection splitting technique such as WISE can offer nearly optimal utilization and minimal transfer time.

References:
[1] J. S. Stadler and J. Gelman, "Performance Enhancement for TCP/IP on a Satellite Channel", MILCOM 98
[2] J. S. Stadler and E. Modiano, *"An On-Board Packet Processing Architecture for the Advanced EHF Satellite System"*, MILCOM '97
[3] W. R. Stevens, *"TCP/IP Illustrated, Volume I"*, Addison-Wesley 1994
[4] IETF RFC 1323 "TCP Extensions for High Performance"
[5] IETF RFC 2001 "TCP Slow Start, Congestion Avoidance, Fast Retransmit, and Fast Recovery Algorithms"
[6] IETF RFC 2018 "TCP Selective Acknowledgement Options"
[7] V. Arora, et. Al., "Asymmetric Internet Access Over Satellite-Terrestrial Networks", AIAA Proc. Of the 16th International Communications Satellite Systems Conference and Exhibit, pp. 476-482, 1996
[8] J. S. Stadler, *"A Link Layer Protocol for Efficient Transmission of TCP/IP via Satellite"*, MILCOM '97

Development and Implementation of an Adaptive Error Correction Coding Scheme for a Full Duplex Communications Channel

John W. Waterston
wdwater@yahoo.com

and

C. Wooten
W. Bennett, T. B. Welch
wmbennet@nadn.navy.mil

Department of Electrical Engineering
United States Naval Academy
Annapolis, MD 21402, USA

Abstract

This paper investigates via simulation, the bit error probability (BEP) associated with a variable redundancy coding scheme operating in a wireless environment. Within a slowly varying(flat fading) Rayleigh channel, this algorithm provides increased throughput over fixed coding implementations. From a family of BCH codes of the same block length (n=63), a code with appropriate redundancy is chosen depending on the receiver's estimation of the current conditions experienced in this channel. Two different decision techniques are compared. The first method statistically evaluates the receiver's input and calculates the signal to noise ratio (Eb/No), while the second method observes the number of corrected errors in recently decoded blocks. Once deciding to modify the correction ability of the code, the decoder transmits the decision to the encoder over a low bandwidth feedback channel, allowing the correction ability to be changed on a block by block basis. This algorithm is implemented in software, and therefore can be optimized for many real world communications systems. The low cost of high speed microprocessors and DSPs allow for the development of a robust adaptive coding system in hardware. The results are compared against fixed coding implementations and show that the adaptive process maintains a better efficiency ($\eta = k/n$) of information rate while keeping the bit error probability near the level obtained by maximum encoding.

1. Introduction

In 1948 Shannon stated that coded data can be transmitted at rates near the channel capacity with an arbitrarily small probability of error by coding. This statement has led to much research in finding the most efficient methods of transmitting data in a noisy environment. Benice in 1966

230

looked at variable redundancy coding and said that this "adaptive technique was of little value." However, the addition of greater computational ability brings new insight to the modern evaluation of this method [1].

In many applications codes are used with fixed redundancy. These codes are set to correct errors in worst-case conditions or optimally chosen for long-term channel conditions. The problem with this technique is that some channels have time varying parameters. When the channel is performing better than the worst-case estimate, the fixed system is utilizing a code with too much redundancy. Instead of transmitting these redundant bits, data could be sent in their place. An adaptive system can take advantage of these changing conditions.

Different types of adaptive techniques to improve the quality of digital transmission include slowing data rates, increasing power at the transmitter, employing diversity, and variable error control with codes [1]. A variable coding system employing BCH error correcting codes was used for this research. Three different adaptive coding methods exist. One method keeps the block length (n) constant and varying the information content (k). Another varies the block length of the code and maintaining the code rate ($R=k/n$), while the final method varies both parameters simultaneously [1,3].

2. Communications System

A variable redundancy coding scheme was decided upon, where the block length would remain constant and the information length would change. The family of BCH codes with length 63 was chosen because it has 11 possible combinations between (n, k). This was few enough to keep the design simple, but allowed for enough states to demonstrate the adapting system. Keeping the block length constant would also keep the transmitter operating at a constant bandwidth, and simplify the receiver structure.

The communications system for this project can be visualized by a multi-stage process which is then implemented in MATLAB® code (Fig.1).

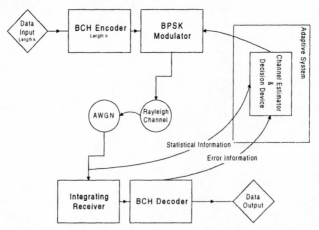

Fig.1 - System Block Diagram

The "data" chosen for simulation is characterized by a random set of 0's and 1's that are equiprobable, but could be replaced by digitized voice signals or satellite telemetry for a real world system. This data is sent to the BCH encoder, which adds the appropriate amount of redundancy forming blocks of length 63. Next, the data is sampled four times for every bit and modulated into a BPSK signal, where -1 represents a binary zero and 1 represents a binary 1. Now the signal is ready for transmission.

The data is sent though a Rayleigh fading channel representing the Doppler fading of a digital cellular phone (IS-95) operating in a moving environment at 9600 bps. This signal is attenuated/amplified by the channel, and also affected by additive white Gaussian noise (AWGN) at the receiver. This channel data is sent to the receiver and also the adaptive system for statistical analysis.

The receiver integrates over every four samples (a bit) and determines whether a 0 or a 1 was sent. Perfect framing and synchronization were assumed for the simulations. This data is sent to the BCH decoder where the errors induced by the channel are removed. Information about the corrected errors is sent to the adaptive system for analysis.

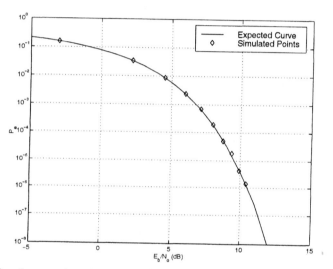

Fig. 2 - BPSK performance in an AWGN channel

To ensure that the simulation of the system was operating properly the output of each subsystem was checked against published equations. This was performed by using a Monte-Carlo method, where successive bits of data are sent through the system until the error probability reaches a statistically significant value. A BPSK signal in AWGN is shown in Figure 2. The expected curve follows the function below. [7]

$$P_e = \text{erfc}\left(\sqrt{\frac{E_b}{N_o}}\right)$$

232

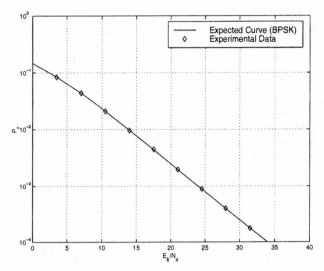

Fig. 3 - Rayleigh Channel and AWGN

This was followed by an AWGN and a Rayleigh channel (Fig. 3) The function has the form for a signal diversity of 1 (L=1) [5].

$$p_r = \left[\tfrac{1}{2}(1-\mu)\right]^L \sum_{k=0}^{L-1} \binom{L-1+k}{k} \left[\tfrac{1}{2}(1+\mu)\right]^k$$

$$\mu = \sqrt{\frac{\overline{\gamma}_c}{1+\overline{\gamma}_c}} \qquad \overline{\gamma}_c = \text{ mean SNR}$$

Finally the BCH error curves were plotted (Fig. 4) with published upper bounds for

$$P_h < P_{ub} = \tfrac{1}{n} \sum_{i=e+1}^{n} \binom{n}{i}(i+e)p^i(1-p)^{n-1}$$

$$e = \frac{(d-1)}{2} \qquad p = \text{erfc}\left(\sqrt{\frac{2k}{n}\frac{E_b}{N_o}}\right) \quad n = \text{block length } d = \text{minimum distance}$$

comparison [6,8]. Note how the experimental data approaches the bounds at lower error rates.

These three figures (Figs. 2 - 4) show that each system is properly implemented and outputting valid simulation results.

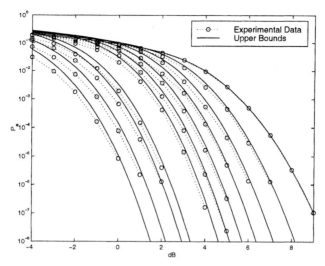

Fig. 4 - Family of BCH codes length 63.

3. Adaptive System

The adaptive system involves all of the decision processes made at the receiving side of the system. Two different methods are examined. The statistical method of estimating the channel looks at the incoming signal and performs statistical calculations to find the mean and standard deviation of the signal which would represent the attenuation and the noise voltage. This method from Cress solves for the signal level represented by the location of the 0 and 1 peaks in the probability distribution function (PDF) using the 2nd and 4th moments [2].

$$E_{2,4} = \int_{-\infty}^{\infty} \left(x - m_x\right)^{2,4} p(x)dx$$

$$\hat{m} = \tfrac{3}{2}\sqrt[4]{E_2^2 - \tfrac{1}{2}E_4}$$

Figure 5 shows the Rayleigh channel and the estimation given by the system at the receiver. Notice how the estimate deviates from the actual value during periods of high attenuation.

In order to adapt the coding in the this first method, the system takes this estimated value of the channel conditions and enters a look up table. This table lists what code is appropriate for the estimated conditions. This look up table is generated by noting the dB level at which the error

234

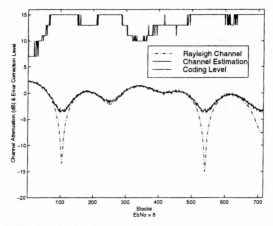

Fig. 5 – Statistical Estimation Method

curves cross a line of constant BEP (10^{-4} in the simulation) in Fig. 4.

The second method, shown in Figure 6, uses the corrected error information provided as an output of the decoder. This value varies with the attenuation of the Rayleigh channel. Instead of calculating a value describing the current channel, the decision device looks at the errors corrected within the current block and compares this number to the current code's maximum correction capability. If the decoder corrected more than the upper threshold (75% of the maximum capability) for the current code (eg. 5 errors when the code can only correct 6), the decision device will shift to a code with more redundancy. Likewise if the number of errors is below the lower threshold (25% of the maximum capability) the system will decrease the next code's redundancy.

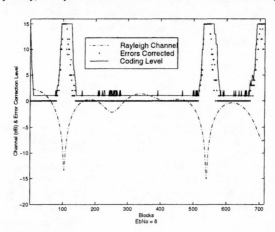

Fig. 6 - Error Evaluation Method

If small bursts of noise are creating excessive level changes, a smoothing factor can be added, so that decisions are made from an average of multiple blocks. This reduction of level changes reduces the load on feedback channel, which may be limited by design constraints.

In both figures, the upper solid line shows the maximum correction capability of the code used for each block. This changing level is an indicator of the working adaptive process. Higher levels of coding reduce the code's efficiency. Note the large difference in coding efficiency between these two different methods.

To communicate each adaption decision, a message is sent via a low bandwidth feedback channel to the encoder. During simulation this feedback channel was assumed to be error free and without delay. This allowed necessary information about coding changes to implemented before the next block was coded.

4. Performance

Initially the first simulations of the system were using a non-realistic channel operating with higher gains. For this type of scenario, the first method (statistical estimation) performed the best, but once a realistic Rayleigh channel was implemented, the estimator's loss in performance forced the use of a second method. These two methods are contrasted during simulation.

The criteria that they are compared with include efficiency, BEP, and a performance metric. Efficiency η is the total amount of information bits transmitted divided by the total number of bits in blocks (total number of blocks times 63). The bit error probability (BEP) is the number of information bits received in error after decoding and correction, divided by the total number of bits of transmitted information. The performance metric allows both parameters to be combined and analyzed simultaneously.

$$\text{Metric} = -\log_{10}(\text{BEP}) \cdot \eta\%$$

Figures 7 - 9 show the performance of both types of adaptive systems while changing the level of the AWGN encountered at the receiver. Examples of the minimum and maximum levels of fixed coding are shown for reference.

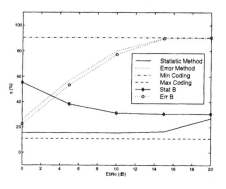

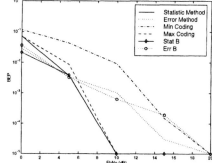

Fig. 7 - Efficiency (η) vs. Eb/No

Fig. 8 - BEP vs. Eb/No

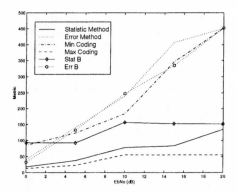

Fig. 9 - Metric vs. Eb/No

The BCH codes used for this simulation have a range of Eb/No in which they operate most efficiently. For the length 63 family of BCH codes this range extends upward from -0.7 dB at a BEP of 10^{-4}. The code would perform better if it could extend this range into lower dB.

The major problem encountered was that the BEP at low Eb/No was unacceptable, often approaching 10^{-1}. Rayleigh attenuation in addition to the large amounts of AWGN creates a quickly changing environment. The adaptive system does not provide any benefit when encountering such conditions.

Two types of errors exist in this simulation, threshold errors and transition errors. Threshold errors occur when too many errors occur for the maximum level of redundancy, this often occurs below -5dB. The inability of the system to quickly adjust creates a transition error, this leaves the current code with too little correction ability for the errors encountered. Both codes have a limitation in how quickly they can react to a changing channel. They both only change one coding level at a time. This is why the simulation is designed for a slowly varying channels.

The Type B systems shown in the three figures above, are attempts to improve the performance of the original adaptive systems. For "Stat B" the different levels in the look-up table were all shifted by a constant of -2.5 dB. This was an attempt to increase the system's efficiency and metric. In "Error B" the upper threshold level was moved from 75% to 50% of the maximum correction ability. This was an attempt to improve the adaption response and increase the system's BEP.

5. Conclusions

Overall, both adaptive methods show benefits compared to fixed coding, but the metric demonstrates the error method to be superior. At 10 db this adaptive system has 29% gain over maximum encoding.

The benefits to this error based decision method include simplicity and ruggedness. The adaption decision is made from existing BCH decoder outputs and does not rely on further signal analysis, making the system more simple. The statistical method showed considerable problems creating a channel estimate in severe conditions. The error method is rugged since severe noise and attenuation do not affect the system's decision making ability. The current problem with the error system is its lag in responding to quick changes of the channel which lead to transition errors.

The two Type B systems prove that the performance of the original designs can be improved. In future investigations performance needs to be maximized for all levels of Eb/No.

The low cost of modern microcomputers and DSPs to perform the tasks of coding and making decisions, could make the computational overhead associated with such a system worthwhile. By using other types of codes (Reed Solomon, convolutional or turbo), the system may gain an increase in performance. Also an ARQ system could be implemented when the capability of the BCH code is exceeded. This would ensure a much lower probability of error than shown in this simulation. With these improvements this adaptive system could be incorporated into modern communications systems. Two such modern systems include a satellite link with rain attenuation or ship-to-ship data links in the Navy [4].

References

[1] Benice, R. J. "Adaptive Modulation and Error Control Techniques." IBM Corporation, 1966.

[2] Cress, D. E. and Ebel, W. J. "Turbo Code Implementation Issues for Low Latency, Low Power Applications." Proceedings of the MPRG Symposium on Wireless Personal Communications, June 10-12, 1998.

[3] Farrell, P.G. "Coding for Noisy Data Links.' Ph. D. dissertation. University of Cambridge, 1969.

[4] Ha, Tre. Digital Satellite Communications. New York: Macmillan Publishing Company, 1986.

[5] Proakis, John G. Digital Communications. Second Edition. McGraw-Hill Book Company, New York, 1989.

[6] Sklar, Bernard. Digital Communications: Fundamentals and Applications. Englewood Cliffs: Prentice Hall, 1988.

[7] Stremler, Ferrel G. Communications Systems. Third Edition. New York: Addison-Wesley Publishing Company, 1990.

[8] Weng, L.G. "Soft and Hard Decoding Performance Comparisons for BCH Codes," Proceedings of the International Conference on Communications, 1979.

Simulink Simulation of a Direct Sequence Spread Spectrum Differential Phase Shift Keying SAW Correlator

S. M. Nabritt, M. Qahwash, M.A. Belkerdid

Electrical and Comp. Engr. Dept, University of Central Florida, Orlando FL 32816
smn12713@pegasus.cc.ucf.edu

ABSTRACT

This paper presents the simulation results of a differential phase shift keying (DPSK) single SAW based correlator for direct sequence spread spectrum applications. The DPSK modulation format allows for noncoherent data demodulation while the SAW device correlator acts as the despreader. The simulator will be using two parallel correlators and a one data bit delay element, while the saw based system uses two in-line correlators. When implemented on SAW devices, this in-line structure has the advantage of an inherent one data bit delay, lower insertion loss, and less signal distortion than the parallel structure. The DPSK correlator was fabricated first on LiTaO₃ substrate, and on a (100) cut GaAs substrate with SAW propagation in the <110> direction. The device autocorrelation function was measured from the LiTaO₃, and the peak to sidelobe ratio was somewhat lower than expected, and the system performance was not measured. The GaAs system offered good results in terms of despreading (autocorrelation) and in terms of data demodulation. This simulation generates autocorrelation predictions as well as data demodulation. This paper presents computer simulation predictions and compares them to experimental results from the devices built on the GaAs substrate. The computer simulation predictions were in good agreement with experimental results.

INTRODUCTION

Direct sequence spread spectrum DSSS) has become the modulation method of choice for wireless local area networks (WLAN's), and personal communication systems (PCS), because of its numerous advantages, such as jammer suppression, code division multiple access (CDMA), and ease of implementation. Spread spectrum systems are most favorable for indoor communication needs [1] and digital radio links [2], where most of the applications are found.

This work was funded in part by a grant from the Florida Department of Transportation, Contract N0 BB-534

The DPSK modulation format allows noncoherent data demodulation while the SAW device correlator acts as the despreader. The simulink simulation was modeled using the conventional technique of two parallel correlators with a one data bit delay element. Where as the SAW correlator used two in-line correlators.

This paper presents the simulation of a DSSS DPSK data modulator and despreader. The results from the simulations are then compared to the experimental data obtained from the SAW DSSS DPSK data modulator and despreader fabricated on GaAs substrate.

DPSK TRANSMITTER SIMULATION

When differentially encoding an incoming message, each input data bit must be delayed until the next one arrives. The delayed data bit is then mixed with the next incoming data bit. The output of the mixer gives the difference of the incoming data bit and the delayed data bit [2]. The differentially encoded. data is then spread by a high-speed pseudonoise sequence (PN). This spreading process assigns each data bit its own unique code, allowing only a receiver with the same spreading sequence to despread the encoded data [3]. The 63-bit pseudonoise sequences (PN) used in this papers are generated by a 6th order maximal length sequence shown in equation one.

$$g(x) = x^6 + x^5 + 1 \tag{1}$$

The implementation of the 6th order polynomial using d flip-flops is shown in Figure 1.

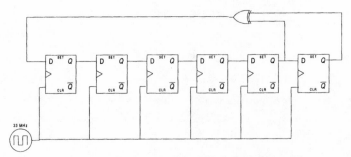

Figure 1. Implementation of 6th Order Polynomial

The maximal length spreading sequence uses a much wider bandwidth than the encoded data bit stream, which causes the spread sequence to have a much lower power spectral density [2]. A baseband DPSK transmitter is depicted in Figure 2.

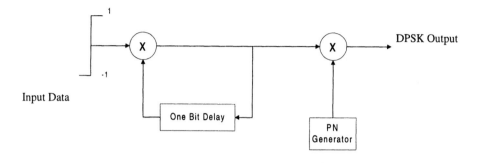

Figure 2. DPSK Encoder Model

The transmitted signal is then given by

$$x(t) = m(t)\ c(t) \tag{2}$$

Where m (t) is the differentially encoded data, and c(t) is the 63 chip PN spreading code [4]. The next step was to simulate the DPSK DSSS transmitter using Simulink. The designed model for the transmitter is shown in Figure 3. Figure 3 consists of a hierarchical system where blocks represent subsystems and oscilloscopes are placed along the path for display purposes.

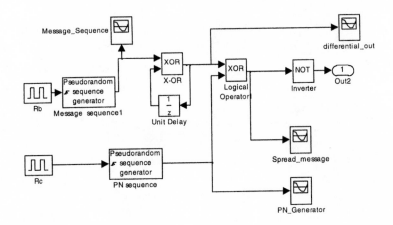

Figure 3. Simulink Model of DPSK DSSS Transmitter

The message data, its differentially encoded version, and the differentially spread waveforms are displayed in Figure 4.

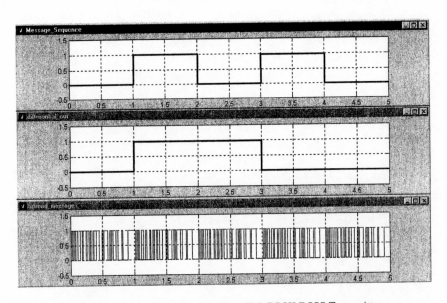

Figure 4. Output Waveforms of Simulink DPSK DSSS Transmitter

DPSK RECEIVER SIMULATION

To demodulate the message the incoming signal is split into two parallel paths [2]. The two paths are then fed into two identical matched filters with the input to one having a delay of 63 chips. The outputs of the two matched filters are denoted by x_1 (t) and x_2 (t) and are given by

$$x_1(t) = d(t\text{-}t_0) R_c(t) \qquad (3)$$

$$x_2(t) = d(t - t_0 - T_b) R_c (t - T_b) \qquad (4)$$

where T_b is the data bit period, and $R_c(t)$ is the autocorrelation function of the 63 chip pseudorandom sequence. Since there are exactly 63 chips per data bit the PN sequence is periodic with T_b so

$$R(t) = R_c(t - T_b) \qquad (5)$$

The two outputs of the matched filters are then mixed and then low pass filtered and the original message is recovered. A block diagram for this receiver is shown in Figure 5. This design was then modeled using Simulink. A detailed Simulink system is shown in Figure 6.

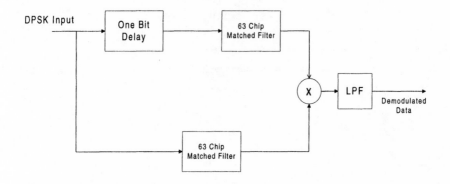

Figure 5. Block Diagram of DPSK DSSS Receiver

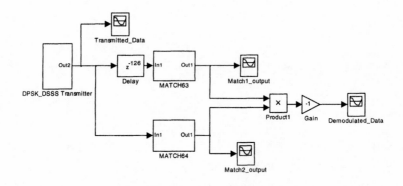

Figure 6. Simulink Model of DPSK DSSS Receiver

The simulation results of the DPSK receiver are depicted in Figure 7 and 8. Figure 7 displays the data, encoded data, encoded spread data, the autocorrelation of the first 63-chip FIR filter, and the autocorrelation of the second 63-chip FIR filter [6]. Figure 8 displays the data, differentially encoded data, and the demodulated data.

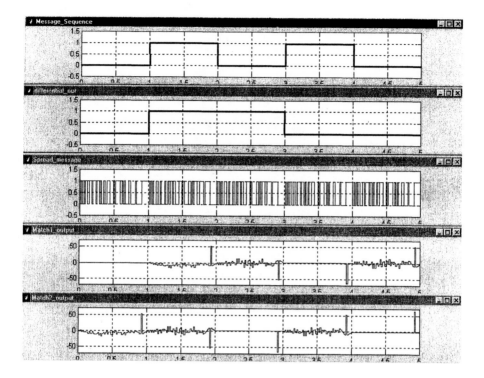

Figure 7. Output Waveforms for Simulink DPSK DSSS Receiver

246

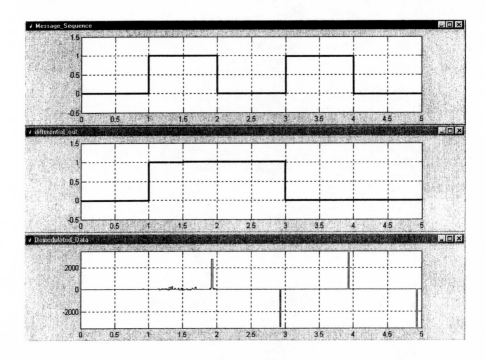

Figure 8.   Output Waveforms for Simulink DPSK DSSS Receiver

SAW IMPLEMENTATION

The implementation of the SAW DPSK DSSS can be implemented using the conventional method shown in Figure 10.   Haartsen [5] proposed a SAW device design which combines all the necessary functions into one SAW device in order to overcome problems such as high insertion loss, limited bandwidth, and temperature effects which are encountered when using three separate SAW devices, (one SAW device for the match filters for each of the parallel paths and one SAW device to generate the one bit delay for one of the parallel branches). The layout of this new structure involving the use of two in-line-coded transducers driven by a wideband input transducer totally eliminates the need for a delay line [6].   The reduction in the number of SAW devices dramatically reduces the insertion loss while also reducing packaging requirements and packaging parasitics. Each matched filter is 63 chips long which generates a one data bit time delay between the output of the two in-line coded matched filters. By structuring all functions of the DSSS DPSK receiver in one device the temperature effects will also be reduced [6]. Figure 9 depicts the structure of the in-line SAW based DSSS DPSK receiver.

The SAW in-line correlators were built on {100} GaAs with the SAW propagating in the <110> direction [7]. Figure 10 shows the principal layout of the SAW in-line correlator structure and Table 1 contains the design specifications. A chip length of 8 wavelengths was used and the chip pattern was repeated 63 times. The 63 chip m-sequence PN code was applied to each chip by changing the center of transduction to the appropriate weight, either "+1" or "-1", thus providing the positive and negative polarity required for a binary PN code. Gratings which were used at both ends of the SAW devices in order to reduce electromagnetic feed-through were 64 wavelengths long and the input transducers were 8 wavelengths long [8]. Each structure was laid out adjacent to the next one without any extra spacing. The total SAW device length was 25.28 mm.

Wavelength (um)	20
Center frequency (MHz)	142.8
Electrode width (um)	2.5
Beam aperture (wavelengths)	90
Metal	Cr/Au
Metal thickness (^{A})	20/600

Table 1. Experimental SAW device in-line correlator design parameters.

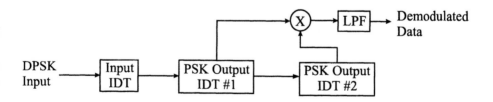

Figure 9. Conventional SAW Implementation on DPSK DSSS Receiver

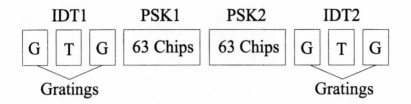

Figure 10. In-Line SAW Implementation of DPSK DSSS Receiver

EXPERIMENTAL RESULTS

An RF The DPSK signal was generated and fed to the GaAs saw correlator. An elaborate RF test bed was set up and the outputs of both SAW filters were measured. These outputs display the autocorrelation of the modulated, differential encode and spread waveforms. The autocorrelation of the two 63 chip SAW filters are depicted in Figure 11.

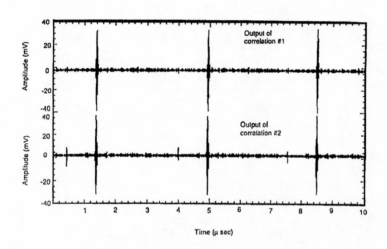

Figure 11. Outputs of 63-Chip SAW Filters

CONCLUSION

This paper presented the simulation results of a SAW based Correlator/demodulator of a direct sequence spread spectrum differential Phase shift keying system. Two 63 – chip FIR filters were used as matched filters in both arms of the DPSK receiver. The simulation results, in terms of autocorrelation functions of the matched filter agreed well with experimental results obtained with a SAW correlator receiver built ona GaAs substrate.

REFERENCES

[1] K. Tsubouchi, H. Nakase, A. Namba, K. Masu, "Full Duplex Transmission Operation of a 2.45 GHz Asynchronous Spread Spectrum Modem Using a SAW Convolver," *IEEE Trans. UFFC,* Vol. 40, No. 5, Sept. 1993, pp. 478-482.

[2] M. Kavehrad, G.E. Bodeep, "Design and experimental results for a direct-sequence for spread-spectrum radio using DPSK modulation in indoor, wireless communications," *IEEE J. SAC,* Vol. 5, 1987, pp. 815-823.

[3] R.E. Ziemer, R.L. Peterson, "Digital Communications and Spread Spectrum Systems," *Macmillan Publishing Company, New York,* 1985, p. 213-222, 569ff, 750.

[4] G.L. Turin, "Introduction to spread-spectrum antimultipath techniques and their application to urban digital radio," *Proc. IEEE,* Vol. 68, 1980, pp. 328-353.

[5] J.C. Haartsen, "A Differential-Delay SAW Correlator for Combined DSSS Despreading and DPSK Demodulation," *IEEE Trans. on Commun.,* Vol. 41, No. 9, Sept. 1993, pp. 1278-1280.

[6] F. Moeller, A. Rabah, S.M. Richie,M.A. Belkerdid, D.C. Malocha, "Differential Phase Shift Keying Direct Sequence Spread Spectrum Single SAW Based Correlator Receiver," *IEEE Ultrason. Symp. Proc.,* pp. 189-193, 1994.

[7] J. Enderlein, S. Makarov, E. Chilla, H.-J. Froehlich, "Mass sensitivity of temperature stabilized surface acoustic wave delay lines on GaAs," *Sensors and Actuators B,* 24-25, pp. 65-68, 1995.

[8] R.T. Webster and P.E. Carr, "Rayleigh waves on GaAs, Rayleigh wave theory and application," *Springer Verlag, Berlin,* 1985, Ch. 3, pp. 123-130.

INDEX